Environment, Media and Communication

Media and communication processes are central to how we come to know about and make sense of our environment and to the ways in which environmental concerns are generated, elaborated, manipulated and contested. The second edition of *Environment, Media and Communication* builds on the first edition's framework for analysing and understanding media and communication roles in the politics of the environment. It draws on the significant and continuing growth and advances in the field of environmental communication research to show the increasing diversification and complexity of environmental communication. The book highlights the persistent urgency of analysing and understanding how communication about the environment is being influenced and manipulated, with implications for how and indeed whether environmental challenges are being addressed and dealt with.

Since the first edition, changes in media organisations, news media and environmental journalism have continued apace, but – perhaps more significantly – the media technologies and the media and communications landscape have evolved profoundly with the continued rise of digital and social media. Such changes have gone hand in hand with, and often facilitated, enabled and enhanced shifting balances of power in the politics of the environment. There is thus a greater need than ever to analyse and understand the roles of mediated public communication about the environment, and to ask critical questions about who/what benefits and who/what is adversely affected by such processes.

This book will be of interest to students in media/communication studies, geography, environmental studies, political science and sociology as well as to environmental professionals and activists.

Anders Hansen is Associate Professor in the School of Media, Communication and Sociology, University of Leicester, UK. He is Associate Editor of *Environmental Communication*; Founder and immediate-past Chair of the IAMCR Group on Environment, Science and Risk Communication; founding member, and Executive Board Member and Secretary 2011–2015, of the International Environmental Communication Association (IECA).

Routledge Introductions to Environment Series

Series Editor:
Timothy Doyle

Environmental Science texts

Atmospheric Processes and Systems
Natural Environmental Change
Environmental Biology
Using Statistics to Understand the Environment
Environmental Physics
Environmental Chemistry
Biodiversity and Conservation, 2nd Edition
Ecosystems, 2nd Edition
Coastal Systems, 2nd Edition

Environment and Society texts

Environment and Philosophy
Energy, Society and Environment, 2nd Edition
Gender and Environment
Environment and Business
Environment and Law
Environment and Society
Representing the Environment
Environment and Social Theory, 2nd Edition
Environmental Values
Environment and Tourism, 2nd Edition
Environment and the City
Environmental Policy, 2nd Edition
Environment and Economy
Environment and Food
Environmental Governance
Environment and Citizenship
Sustainable Development, 2nd Edition
Environment and Politics, 4th Edition
Environment and Tourism, 3rd Edition
Environment, Media and Communication, 2nd Edition

Environment, Media and Communication

Second edition

Anders Hansen

LONDON AND NEW YORK

Second edition published 2019
by Routledge
2 Park Square, Milton Park, Abingdon, Oxon, OX14 4RN

and by Routledge
711 Third Avenue, New York, NY 10017

Routledge is an imprint of the Taylor & Francis Group, an informa business

First edition published by Routledge 2010

British Library Cataloguing-in-Publication Data
A catalogue record for this book is available from the British Library

Library of Congress Cataloging-in-Publication Data
Names: Hansen, Anders, 1957– author.
Title: Environment, media and communication / Anders Hansen.
Description: 2nd edition. | Milton Park, Abingdon, Oxon; New York, NY: Routledge, 2019. | Series: Routledge Introductions to Environment Series: Environment and society texts | Previous editon: 2010. | Includes bibliographical references and index.
Identifiers: LCCN 2018018294| ISBN 9781138650459 (hbk) | ISBN 9781138650473 (pbk) | ISBN 9781315625317 (ebk)
Subjects: LCSH: Mass media and the environment.
Classification: LCC P96.E57 H36 2019 | DDC 070.4/493337—dc23
LC record available at https://lccn.loc.gov/2018018294

ISBN: 978-1-138-65045-9 (hbk)
ISBN: 978-1-138-65047-3 (pbk)
ISBN: 978-1-315-62531-7 (ebk)

Typeset in Times New Roman
by codeMantra

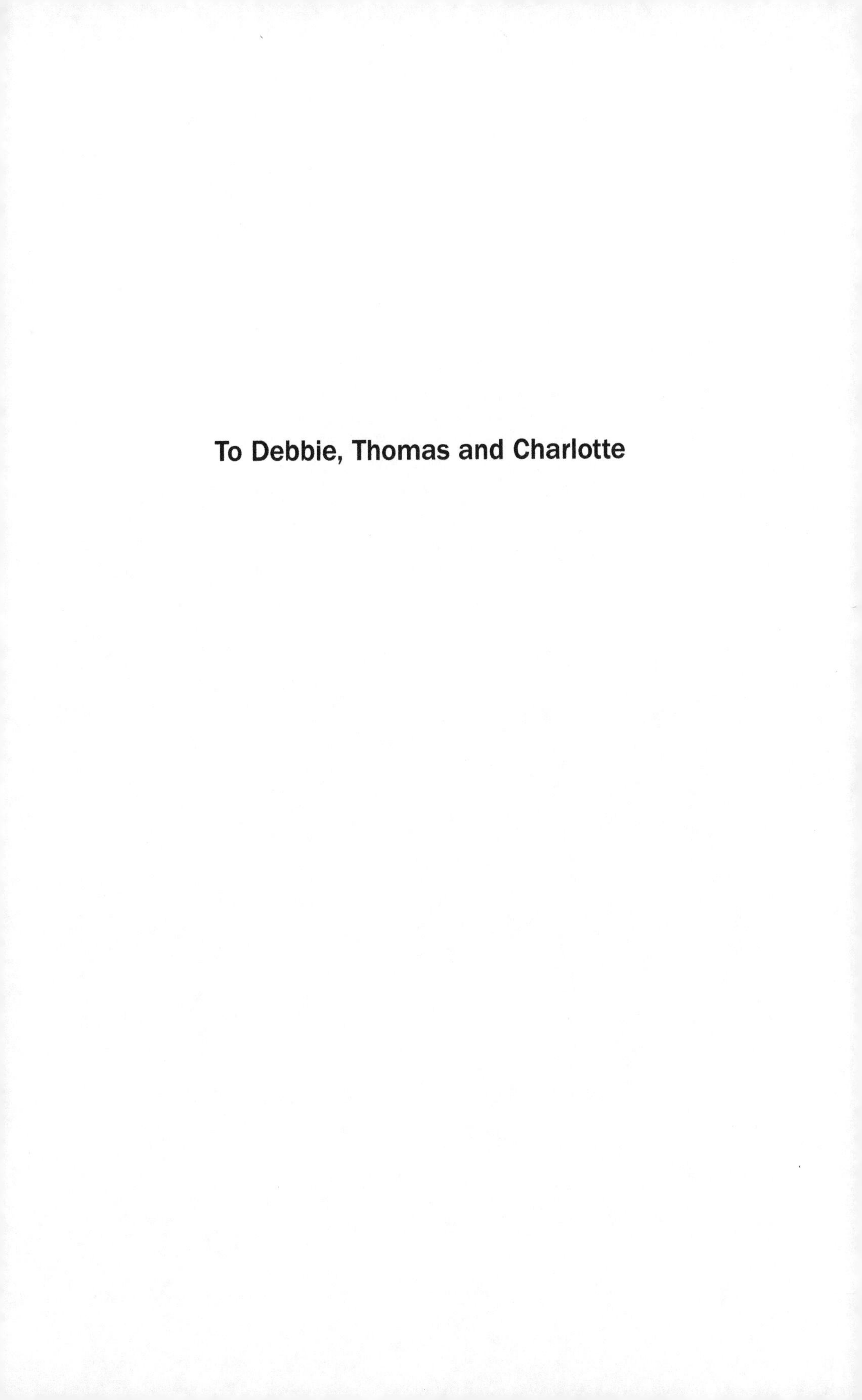

To Debbie, Thomas and Charlotte

Contents

Boxes

Exercises

Preface to Routledge Introductions to Environment Series

Series Editor: Timothy Doyle,

Professor of Politics and International Studies, University of Adelaide, Australia;
Distinguished Research Fellow, Curtin University, Australia;
Chair, Indian Ocean Rim Association Academic Group, Port Louis, Mauritius;
Chief Editor, *Journal of the Indian Ocean Region*, Routledge, Taylor and Francis, London, UK.

It is openly acknowledged that a critical understanding of socio-economic, political and cultural processes and structures is central in understanding environmental problems and establishing alternative modes of equitable development. As a consequence, the maturing of environmentalism has been marked by prolific scholarship in the social sciences and humanities, exploring the complexity of society–environment relationships.

This series builds on the work of the founding series editor, David Pepper, to continue to provide an understanding of the central socio-economic, political and cultural processes relating to environmental studies; providing an interdisciplinary perspective to core environmental issues. David initiated the series by celebrating the close connections between the academic traditions of environmental studies with the emergence of the green movement itself. Central to the goals of the movement were social and environmental change. As the 'new science' of ecology was inter-disciplinary, seeking to understand relationships within and between ecosystems, so too was the belief within the academy (informed by the movement), that real environmental change could only emerge if traditional borders and boundaries of knowledge and power were bypassed, transgressed and, where necessary, challenged.

This bid for engaged knowledge and inter-disciplinarity also informs the structure and 'pitch' of these books. For it is no good communicating with just one particular group within society. It is equally important to construct forms of knowledge which can cross-over demographic and market borders, bringing together communities of people who may never 'meet' in the usual course of events. So, the epistemological design of this series is oriented around three particular audiences, providing an unparalleled interdisciplinary perspective on key areas of environmental study: 1) students (at undergraduate and coursework post-graduate levels); 2) policy practitioners (in civil society, governments and corporations); and 3) researchers. It is important to note, therefore, that these books—though strongly used in diverse levels of tertiary teaching— are also built, in large part, on the primary and often ground-breaking research interests of the authors.

In his own ground-breaking work, David Pepper was particularly interested in exploring the relationships between capitalism, socialism and the environment. David argued that the modern environmentalist movement grew at a rapid pace in the last third of the twentieth century. It reflected popular and academic concerns about the local and global degradation of the physical environment which was increasingly being documented by scientists. It soon became clear, however, that reversing such degradation was not merely a technical and managerial matter: merely knowing about environmental problems did not of itself guarantee that governments, businesses or individuals would do anything about them. Since David wrote his last series preface, this focus has continued to be important, but with special permutations as time has worn on. One more recent, key feature of these society–environment relationships has been the clear differentiation between the environmentalisms of the majority worlds (the global South) and environmentalisms of the minority worlds (the more affluent, global North). Wherein environmentalism came to the less affluent world later (in the 1980s), key environmental leadership is now being provided by activists in the South, oriented around a post-colonial environmentalism: with its key issues of human dispossession and survival: water, earth (food security and sustainability), fire (energy), and air (not climate). Much of the focus in environmentalism in the South relates to a critique of capitalism and its big business advocates as being the major perpetrators of severe environmental problems which confront the Earth. In the global North (where the modern movement began in the 1960s), there has been far more emphasis on post-materialism and post-industrialism and, more recently, building a *sustainable capitalism*.

Climate change is now the neo-liberal, cause celebre of this approach, with its heavily-biased focus on market mechanisms and green consumerism as answers to environmental crises. In fact, climate change, in the global North, has now become so powerful and omnipresent that many more affluent-world green activists and academics now comprehend *all* environmental problems within its rubric, its story. Of course, climate change issues will continue to be crucial to the planet's continued existence, but more importantly, it must be acknowledged that in living social movements – like the green movement – issues will come and go; will be re-ordered and re-arranged on the issue attention cycle; be re-badged under different symbols, signs and maps; and new green narratives, issues and stories will emerge. The environment movement, born in the North – and its associated academic studies – will continue to be the foremost global social movement for change for many years to come – if it can continue to truly engage with the global South – utilising these new and revised banners, issues and colours to continually and creatively mark out its territories, constructing versions of environmentalism for all; not just for the few. And it is within these new sites of politics and knowledge where some of the most exciting advances in the relationships between societies and 'nature' will continue to emerge and be celebrated. Much still is to be learned from our universe, the planet Earth, its human and non-human communities.

Tim Doyle,
October 2014

Preface to the Second Edition

There has never been a better or more exciting time than the present to study, engage with and practice environmental communication. Nor has there ever been a greater need for analysing and understanding the roles of media and communication in the politics of the environment. While research on environment, media and communication has been around for over half a century, the pace of change and development in the media and communications landscape, in mediated communication about the environment, and indeed in scholarly research on environmental communication, has picked up significantly, particularly in the second decade or so of our present century. Public and political debate and action regarding the environment have likewise evolved considerably in recent times with profound implications for how and whether environmental problems and challenges are addressed and acted upon.

Building on the framework of the first edition for analysing and understanding media and communication roles in the politics of the environment, this second edition draws on the splendid growth and advances in recent times in the field of environmental communication research, to show the increasing diversification and complexity of environmental communication, and to stress the continued urgency of analysing and understanding how communication about the environment is being influenced and manipulated, with implications for how and indeed whether environmental challenges are being addressed and dealt with.

Since the first edition, changes in media organisations, news media and environmental journalism have continued apace, but – perhaps more significantly – the media technologies and the media and communications landscape have evolved profoundly with the continued rise of digital and social media. And such changes have gone hand in hand with, and often facilitated, enabled and enhanced shifting balances of power in the politics of the environment. There is thus a greater need than ever to analyse and understand the roles of mediated public communication about the environment, and to ask critical questions about who/what benefits and who/what is adversely affected by such processes.

Working on the second edition has been exciting precisely because of the continued and considerable changes in the media and communications landscape and their implications for how communication, campaigning and policy-making with regard to the environment unfold; and it has been exciting because of the splendid and tremendous growth in environmental communication research and practice.

I am very grateful to Andrew Mould, my editor at Routledge, for suggesting and inviting a second edition, and for his splendid support throughout. Thank you also to editorial assistant Egle Zigaite and colleagues at Routledge for their support and assistance, and to the anonymous reviewers for their constructive feedback.

Many more people than can be listed here have provided important inspiration, ideas, feedback, support and/or friendship during my continued journey with environmental communication. In addition to colleagues acknowledged in the first edition, I would like to acknowledge the inspiration from colleagues in the International Environmental Communication Association (IECA – https://theieca.org/); in the IAMCR Environment, Science and Risk Communication group (https://iamcr.org/s-wg/working-group/esr); my co-editor of the *Palgrave Studies in Media and Environmental Communication* series, Steve Depoe; my co-editor of the *Routledge Handbook of Environment and Communication*, Robert Cox. I am also grateful to my 'environmental communication' colleagues at the University of Leicester, for their continued inspiration, collaboration and particularly for their magnificent support in the organisation and hosting of the IECA biennial Conference on Communication and Environment at the University of Leicester in 2017 (https://theieca.org/conference/coce-2017-leicester). Chapter 8 draws from and elaborates on Hansen, A. (2017a). Media representation: environment. In P. Rössler (Ed.), *The International Encyclopedia of Media Effects*: John Wiley & Sons, Inc., doi: 10.1002/9781118783764.wbieme0148; and on Hansen, A. (2016). The changing uses of accuracy in science communication. *Public Understanding of Science, 25*(7), 760–774. doi: 10.1177/0963662516636303.

1 Introduction

The environment, environmental, climate change, global warming, greenhouse effect, acidification, ozone depletion, bleaching, micro-beads, plastics, species extinction, carbon footprint, carbon offsetting, fracking, nuclear, Greenpeace, deforestation, soil erosion, flooding, desertification, pollution, ecology, organic foods, GM crops, GMOs, pesticides, toxic waste, landfill, sustainable development, sustainability, eco-[anything] and green-[anything] … these are terms which are – or have become – familiar parts of our everyday public vocabulary. And like the term 'the environment' itself, they have come to be associated with a particular public conversation about 'problems' to do with our relationship with our natural environment. They have come to be associated with what we might call an environmental discourse, which, in its particular form and with its particular view of the world, is itself of relatively recent origin, dating back only to the early 1960s. Indeed, many see as an important symbolic starting point of a new environmental perspective – distinct from, for example, earlier conservation perspectives – the publication in 1962 of American biologist Rachel Carson's evocatively titled book *Silent Spring*.

There are a couple of significant things to note about the public vocabulary on the environment and environmental change: First, it is a vocabulary which is familiar and recognisable, but probably, for most, only in a superficial kind of way; that is, behind most of these terms lurk some immensely complex issues, that require a great deal of engagement – scientific, philosophical, ethical, moral, economic, and so on – well beyond what most of us have the time, or perhaps inclination, to delve into. Second, it is very much a time/history-bound vocabulary in the sense that it fades in and out of public focus *and* in the sense that parts of it, including the meanings and connotations associated with particular terms, change and evolve over time.

Furthermore, it is of course not merely a linguistic or word-based vocabulary, but very much (in the digital media environment) a *visual* and indeed fully multi-modal communications vocabulary. Thus, as with the words/terms listed above at the start, images of polar bears on ice floes, melting/calving/breaking glaciers, logging, the rainforest, cooling towers and chimneys, industry-scapes, floods and hurricanes, the globe/planet Earth, deserts, large uniform fields of maize or wheat, and so on, not only seem perfectly familiar but are also – whether consciously or subconsciously – instantly recognisable as 'belonging' to the (present) visual vocabulary of environmental discourse, rather than, for example, to the discourses of travel brochures, history textbooks, geography or other science textbooks, and so on.

But visual images used in communication about the environment do not acquire their 'meaning' (such as 'environmental problem' or 'climate change') by themselves, nor indeed – despite their seeming photographic 'window on reality' quality – do they inherently or intrinsically carry a particular meaning or merely 'represent' what they ostensibly show. The elevation to iconic or representative status, and the public identification of these images as belonging to a particular discourse, requires visual signification 'work' in much the same way as terms such as 'climate change' and 'carbon footprint' only become meaningful to us through repeated explanation and association.

Like word definitions, the meaning of pictures or images is anything but static; the elevation of particular images to 'iconic' – in the sense of 'representative' – status as images *representing* a particular meaning, such as 'climate change' or 'environmental devastation' or, perhaps more obliquely, 'threatened environments' is an ongoing process drawing on what Linder (2006: 129–130) aptly refers to as 'an extensive collection of semiotic resources' and involving 'a substantial amount of appropriation and pastiche between them, as they exploit newly established signs in novel variations'.

A key characteristic of the building of a public (visual) vocabulary of the environment and environmental issues is the *abstraction* or *de-contextualisation* of images *from* specific identifiable geographic or cultural environments *to* generic, iconic or 'representative' global environments (Hansen & Machin, 2008). The main point of relevance to the discussions in this book is the point that the meanings and significance which we come to associate with the key terms of the public word and image vocabularies on the environment are the result of a great deal of active – and in many cases highly deliberate – signification and communicative 'work'. Images of melting ice, ice floes, Arctic/Antarctic landscapes, glaciers, rainforests, coral reefs, deserts, and so on, become synonymous with – come to mean or signify – 'threatened environments' and ultimately 'global warming' or 'climate change'. In the past they would have signified something quite different such as 'challenge' or a test of human endeavour and perseverance or indeed simply 'pristine' and aesthetically pleasing environments, as yet untouched and unspoilt by humans.

While the roles of formal education in acquainting us with the public word and image vocabulary of the environment should not be overlooked, most of what we learn and know about 'the environment', we know through mediated communication. Indeed, this applies not only to our beliefs and knowledge about those aspects of the environment, which are regarded as problems or issues for public and political concern, but extends much deeper to the ways in which we, as individuals, citizens, cultures and societies experience, view, perceive and value nature and the natural environment.

Historically, such views have oscillated between the two extremes of, on the one hand, a utilitarian perspective which sees the natural environment as a hostile domain to be controlled and exploited in the name of progress, and, on the other hand, a more romantic view of the natural environment as fragile, pure, pristine beauty in need of protection. Our views of the environment have been, and are being, articulated and shaped through multiple media and forms of communication: paintings, architecture, poetry, literature, film, music, posters, the press, broadcast and other electronic media in the increasingly pervasive digital media environment.

What particularly distinguishes the history of the recent half-century or so is the crucial role played by media and communication in not only helping to define 'the environment' as a concept and domain, but more particularly in bringing environmental issues and problems to public and political attention. Thus, since the emergence and rise of the modern environmental movement in the 1960s, public communications media have been a central public arena for publicising environmental issues and for contesting claims, arguments and opinions about our use and/or protection of the environment. Indeed, a defining feature of many of the most well-known and most politically effective environmental pressure groups has been, and continues to be, their view of both traditional news media and digital and social media as an integral and essential part of their campaigning strategy.

Whereas, in earlier eras, much political decision-making with regard to the environment may have been based largely on expert and scientific evidence/testimony, with a keen eye on economic development and 'progress', such decision-making has increasingly been influenced and governed by how environmental and related issues are presented to and perceived by the public. Communicating about the environment may have been seen in the not so distant past as mainly a matter of making the public understand the science and scientific evidence behind contentious environmental issues. Some of the most controversial environmental issues and debates of the recent period show a very different picture. Whether looking at climate change, energy production, agricultural and fishery policies, animal experimentation or the multiple issues relating to rapid advances in genetic modification and the bio-genetic sciences, it is clear that the battles over these issues are now much more to do with persuasive communication, with 'winning hearts and minds' than they are to do with understanding the 'science' behind these issues.

Communication, then, is a central aspect of how we come to know, and to know about, the environment and environmental issues, and communications media are a central public arena through which we become aware of environmental issues and the way in which they are addressed, contested and, perhaps, resolved.

This book is about how we study and understand the role of communication and media in relation to the environment and environmental issues. It examines the ways in which communication research has contributed, and can contribute, to our understanding of the role of news and other prominent communications media in making the environment and environmental problems issues for public and political concern. A key objective of the book is to draw attention to the highly 'constructed' nature of public communication, conversation and debate about the environment and nature, to show that there is little or nothing that is 'natural' or accidental about the processes by which we as publics come to learn about, understand and interpret environmental issues or problems – indeed, that the mere notion that the environment is an 'issue' or a 'problem' is itself the product of active rhetorical 'work' and construction in the public sphere.

The rise of environmental communication research

Research on media, communication and the environment dates back to the 1970s. Early pioneering studies which have been influential in shaping its development include Anthony Downs's (1972) examination of the public careers of social issues, including the environment as a social problem, and David Sachsman's (1976) study of environmental reporting. At the end of the 1970s came one of the first studies to offer a comprehensive perspective on the key role of the news media in the public construction of the environment as a social problem, namely the study by Schoenfeld, Meier and Griffin (1979). The 1980s saw the publication of important work that in several ways was directly relevant to the rise of media/communication research on the environment: this included work on media and nuclear power (Mazur, 1984; Gamson & Modigliani, 1989), crises/disasters (Nimmo & Combs, 1985; Walters et al., 1989), environmental news journalism (Lowe & Morrison, 1984), and media and science/technology communication (Nelkin, 1987; Friedman et al., 1986).

While the 1970s and 1980s thus produced a steady trickle of media and communication studies relevant to or directly touching on the environment and environmental issues, the 1990s can be characterised as the decade where these trends first coalesced into a distinctive focus on 'media and the environment'. A special issue entitled 'Covering the environment' of the journal *Media, Culture and Society* in 1991 (Corner & Schlesinger, 1991) provided an early thematic focus on media and environment, and this was consolidated further in one of the first academic collections of its kind on *The Mass Media and Environmental Issues* (Hansen, 1993a). This was followed by several book-length introductions with a core focus on media and the environment (Hannigan, 1995; Anderson, 1997; Chapman et al., 1997; Lacey & Longman, 1997; Shanahan & McComas, 1999) as well as others touching directly or indirectly on the key roles of discourse, rhetoric and communication in relation to the environment and nature (Cronon, 1995; Cantrill and Oravec, 1996; Macnaghten and Urry, 1998).

The 1990s consolidation of environmental communication research has continued and become significantly more pronounced during the first two decades of the 21st century. This is evident not just in a marked increase in scholarly research on media, communication and the environment, but in the embedding of environmental communication research within university-level curricula and in sections and groups within national and international communication associations. Indeed, such is the growing interest in this field that new associations such as the International Environmental Communication Association (IECA, launched in 2011) have been formed. Sustaining this trend and the consolidation of 'environmental communication' as a recognisable and recognised field of research is the growing body of book-length publications focused on environmental communication (e.g. Cox, 2006 [now in its 5th edition: Pezzullo & Cox, 2018]; Corbett, 2006; Hannigan, 2006; Boyce & Lewis, 2009; Cottle, 2009; Hansen, 2010; Lester, 2010; Doyle, 2011; Boykoff, 2011; Anderson, 2014; Tong, 2015; Priest, 2016), and the rapid growth in journal articles across a range of science/environment/health and communications journals, including the establishment of academic journals specifically focused on environmental communication, notably the journal *Environmental Communication.* Perhaps the strongest indication of the increasing maturity and consolidation of the field is the emergence of synoptic collections (Hansen, 2014), handbooks (e.g. *The Routledge Handbook of Environment and Communication* (Hansen & Cox, 2015) and the forthcoming *Routledge Handbook of Environmental Journalism* (Sachsman and Valenti, 2019)), encyclopedia (*The Oxford Encyclopedia of Climate Change Communication*; Nisbet, 2018), and book series, such as the *Palgrave Studies in Media and Environmental Communication* launched in 2014 and edited by Hansen and Depoe.

Overview of the book

The chapters which follow draw on the wealth of research now available on environmental communication. The aim is to show some of the complexity of the processes of communicating the environment, to draw attention to the central significance of language/discourse, imagery and cultural values in these processes, and to indicate that communicating about the environment and environmental issues is about a great deal more than just imparting information: it is crucially about power in society, the power to define our relationship with nature and the environment and the power to define (to paraphrase Ryan, 1991) what the 'problem' with the environment is, who is responsible and what course of action needs to be taken.

The book thus addresses questions such as: How far has public-mediated communication, including through the news media, been key to the environment becoming a 'problem' for public and political concern? How is the political process influenced by media coverage of environmental issues? To what extent is public-mediated communication itself influenced or structured by economic pressures, by the professional norms and practices of journalists, by news values and/or by the publicity and news management practices of the major stakeholders in environmental debate (including business, industry, government, environmental pressure groups, scientists, etc.)? Who gets to define what environmental issues are about or how they should be addressed and resolved? How do the media contribute to policing the boundaries of 'acceptable' public debate about the environment? How do mediated images contribute to the formation of public opinion? In what ways do different publics draw on and engage with media representations for making sense of environmental issues? How are nature and references to what is regarded as 'natural' used in media and public debate about controversial issues? How are nature, the natural and invocations of environmental protection or authenticity used in advertising to sell everything from cosmetics and cars to corporate identity?

The book's discussion of these and associated key questions is used to demonstrate the centrality of media and communication in environmental debate. This draws directly from the fast-growing field of environmental communication research, drawing on work encompassing not just the conventional focus on news, but a wider range of media, media genres and forms of communication.

The rest of the book is thus organised as follows: **Chapter 2**, *Communication and the construction of environmental issues*, outlines and discusses a general theoretical framework for understanding and analysing the role of media and communication in relation to environmental issues. It introduces the notion that the environment generally does not 'speak for itself' but that environmental problems only become problems or issues for public concern and political decision-making through claims-making and communication. The chapter introduces and discusses the application of a constructionist perspective to media and environmental communication. It demonstrates how the constructionist focus on competing definitions offers an analytically more productive approach to understanding media roles than more traditional concerns about bias and objectivity in media reporting, and it points to some of the key analytical tools and foci furnished by this approach, including its focus on claims-makers and claims-making, on discourse, on issue careers, on issue resonance, on issue ownership and competition, and so on. The construction of the environment as a social problem is seen as essentially a rhetorical achievement, and the chapter thus points to the centrality in communication analysis of studying the rhetorical idioms, motifs, claims-making styles, frames, settings or public arenas deployed in public discourse about the environment.

Chapter 3, *Making claims and managing news about the environment*, applies the previous chapter's introduction to the constructionist perspective to news and shows that there is little that is 'natural' about environmental news; even 'natural disaster news' can best be understood in terms of something which has to be actively constructed. The chapter focuses on the communication strategies and influence of key stakeholders or interested parties in environmental debate and controversy. How do environmental pressure groups, government departments, scientific establishments, individual scientists/experts, business and industry seek to strategically manage and influence public communication about the environment? How dependent are they on major news media, and how successful are they, or have they been, in 'spinning' environmental stories to their advantage? How successful have they been in influencing public opinion and political decision-making through strategic communication and publicity practices? What are the key ingredients of successful claims-making about the environment?

Relevant criteria of successful campaigning are examined, distinguishing between the key components of raising awareness, claiming legitimacy in the public sphere and invoking action. While much attention in the communications literature has been on the communication and campaigning strategies of environmental pressure groups, the chapter also examines the emerging research literature on the powerful influence and role of corporate communication and image management strategies in the public sphere.

Chapter 4, *The environment as news: news values, news media and journalistic practices*, turns from the previous chapter's emphasis on media-external *sources* and *claims-makers* to the main news media themselves. It focuses on the roles, organisational arrangements, practices and 'communicative work' of the media and media professionals. The chapter discusses how research on news values, on organisational structures and arrangements in media organisations, on the professional values and working practices of journalists and other media professionals can help explain why some environmental issues become news, while others do not; why some environmental issues become issues for media and public/political concern, while others fall by the wayside.

The development of an 'environment beat' and of specialist environmental correspondents is examined, and the chapter discusses whether environmental correspondents are fundamentally different from other types of reporters, or take a different approach to their subject and to their sources than general reporters. How do journalists, reporting on the environment and environmental controversy, deal with the scientific uncertainty which characterises much of environmental debate, how do they secure credibility in their reporting, and how do they deploy traditional journalistic criteria such as objectivity and balance? The chapter surveys the emerging evidence on how environmental journalism and its values and practices are affected by the rise of public relations and related communications intermediaries, by enhanced strategic communication approaches by key stakeholders, by media organisational change, and by the emergence of new – technologically facilitated and enhanced – forms of freelance and citizen journalism. The chapter ends with a discussion of the limitations of the sociology of news framework, and the ways in which some of these limitations have been addressed through perspectives focusing on cultural resonances in the discursive construction of environmental issues.

Chapter 5, *Popular culture, nature and environmental issues*, picks up the suggestion at the end of the previous chapter, that in addition to the traditional sociology of news emphasis on economic pressures, organisational arrangements and professional values/practices, it is necessary – if we wish to understand media coverage of the environment – to take wider cultural resonances and narratives into consideration as well. Chapter 5 thus starts with a focus on the notions of scripts, cultural packages, interpretive packages and cultural resonance, and considers their significance for understanding the nature and potential 'power' of popular media representations of nature and the environment.

The chapter surveys research on the types and origins of images – particularly of science and nature – which have been influential in media and popular constructions of nature and the environment generally, and in media and popular culture constructions of nuclear power, the new genetics and biotechnology, particularly. The discussion in this chapter moves beyond the focus on news coverage of the environment and considers how nature and the environment are constructed ideologically in other media genres, including television entertainment programming and more particularly in wildlife film and television nature programmes. The persistence of key cultural narratives and stories – for example, the Frankenstein story – and the ideological clusters, packages or scripts which are evoked by what are often single trigger words are explored.

The chapter explores the significance of lexis or word choice in media constructions of environmental issues, and the similarly significant contribution of narrative analysis to uncovering the deeper ideological values communicated through wildlife film and nature programming. Drawing on historical studies of selected film genres (e.g. 'science fiction' and 'wildlife/nature films') the chapter discusses how deep-seated cultural narratives have reflected, and in turn shaped, particular ideological interpretations of nature and the environment, including changing dominant interpretations of the environment as either an object of control and exploitation or as something to be protected.

Chapter 6, *Selling 'nature/the natural': advertising, nature, national identity, nostalgia and the environmental image,* explores the promotional use of nature and the environment in advertising and continues the previous chapter's examination of how constructions of nature and the environment change over time. It shows how advertising has been used for promoting environmental messages and awareness, for selling 'green' or 'environment-friendly' products and for improving or promoting the image of large corporations or industries. While explicit environmental appeals and green marketing come and go in advertising, nature imagery and appeals to the natural have been a prominent and relatively consistent feature of commercial advertising for a very long time. The chapter explores how nature/the natural and the environment are constructed and deployed for selling products, and it explores how the uses of nature contribute important boundaries and definitions of appropriate consumption and 'uses' of the natural environment. The uses of nature and environmental images in advertising are examined in relation to the concepts of nostalgia and national/cultural identity, and the chapter investigates variations in images across different cultures and the extent to which such images are either culturally specific or increasingly global/universal.

Chapter 7, *Media, publics, politics and environmental issues*, considers how communication researchers have tackled the perennial 'holy grail' question of how media representations and media coverage of the environment influence public and political perceptions and action.

Ultimately, the assumption, whether explicit or implicit, behind most research into media representations of the environment is that these play a role in shaping and influencing public understanding/opinion and political decision-making in society. This chapter discusses some of the major frameworks and approaches – for example, agenda-setting research, public opinion research, framing analysis and cultivation analysis – which have been used for examining media influence on public understanding, public opinion and political decision-making. It explores the evidence, from these different research approaches and frameworks, for how media representations of environmental issues influence political processes or are interpreted and used by different publics. What evidence, for example, is there from communication research that the media form a significant part in shaping the agenda and nature of public opinion? Do the dynamics of public opinion and media representations in turn influence political decision-making? And how, for rhetorical and political purposes in public debate, are 'the media' and mediated communication themselves constructed/invoked as active players and influencers of public discussion? I conclude by reiterating that the role and influence of media(ted) communication with regard to public/political understanding of the environment are not best examined from the perspective of linear transmission models of communication, but call for appreciation of the increasingly dynamic and interactive nature of communication flows in society, while at the same time retaining the analytical focus on the structured inequality of communication and influence in society.

In the concluding **Chapter 8**, I offer a condensed overview of some of the main emphases and conclusions arising from the book's survey of research on environment, media and communication, and then conclude with some suggestions for the way ahead in environmental communication research. Thus, starting with a brief reiteration of the importance of media and communication in the rise and development of environmental concern, I delineate the disciplinary and theoretical context and some of the main emphases of environmental communication research. Key trends are then summarised under the headings of news media representation of the environment, non-news media representation, and environmental communication in the digital media and communications landscape. I conclude with a summary of the major achievements of environmental communication research to date, and pointing to some of the key challenges ahead, I offer suggestions for where, particularly in light of the rapidly changing nature of the media and communications landscape, future research emphases need to be focused.

2 Communication and the construction of environmental issues

THIS CHAPTER:

- Introduces the constructionist perspective on social problems as a framework for analysing and understanding the role of media and communication in relation to environmental issues.
- Highlights the centrality of claims, claims-makers and the claims-making process in analysis of the emergence, elaboration, and contestation of environmental issues.
- Discusses the construction of social problems as essentially a rhetorical achievement, and points to the analysis of rhetorical idioms, motifs, claims-making styles and settings or public arenas as core components.
- Examines how/whether the constructionist perspective extends to the analysis of natural disasters/accidents.
- Introduces the idea of issue careers and the notion that social problems move through a series of stages in an 'issue-attention' cycle.
- Introduces the concept of framing and discusses how it helps in analysing and understanding media roles in the construction of environmental issues.

Constructing social problems/constructing environmental issues

Why and how should we study media and communication in relation to environmental issues? Perhaps the answer to this emerges from the simple observation that not all environmental problems are publicly recognised as such – as problems requiring some kind of social/political/legislative attention and action – and from the more puzzling observation that environmental issues or problems – over time – fade in and out of public focus in cycles that often seem to have little to do with whether they have been addressed, resolved, averted or ameliorated. Both observations suggest that communication – what is being said about environmental phenomena – is important, and they suggest that a public forum or arena, for example the media, is necessary for environmental phenomena to be recognised as issues for public or political concern.

Environmental issues or problems do not simply emerge and announce themselves as issues requiring a social/political response in the form of legislation, research or a change in public practices and social arrangements. This is not something that is peculiar to environmental problems – similar points have frequently been made in relation to crime, delinquency, poverty, gender and social inequality, racial discrimination, and so on – although, as we shall see in subsequent chapters, there may be some unique aspects of environmental issues that have particular implications for the way in which they come to public attention and become problems for public and political concern.

A key breakthrough in sociology, and one which points directly to the centrality of 'media and communication', was the emergence in the late 1960s and early 1970s of what became known as the constructionist perspective on social problems. The fundamental argument of this perspective was the argument that 'social problems' are not some objective condition in society that can be identified and studied independently of what is being 'said' about it. Problems and issues of various kinds only become recognised as such – as 'problems' or 'issues' – through talk, communication, discourse which defines or 'constructs' them as problems or issues for public and political concern.

One of the first to articulate this perspective was American sociologist Herbert Blumer, who took issue with the way that sociologists had traditionally identified social problems on the basis of public concern. Blumer argued that this was problematic, as many 'ostensibly harmful conditions are not recognised as such by the public, and thus are ignored by sociologists'. Instead, Blumer called for a definition of social problems which recognises these as 'products of a process of collective definition' rather than 'objective conditions and social arrangements' (1971: 298). The key task for research then, according to Blumer, is 'to study the process by which a society comes to recognize its social problems' (1971: 300).

The focus on process and communication evident in the early work of Blumer and fellow American sociologists Malcolm Spector and John Kitsuse received its full articulation in an early article by Spector and Kitsuse (1973) and again in what can appropriately be regarded as the founding book of social constructionism, namely their 1977 book *Constructing Social Problems* (reprinted 1987 and 2000). In it, Spector and Kitsuse define social problems as:

> *the activities of individuals or groups making assertions of grievances and claims with respect to some putative conditions. [...] The central problem for a theory of social problems is to account for the emergence, nature, and maintenance of claims-making and responding activities.*
>
> (2000: 75–76; emphasis in original)

This approach/framework then suggests: a) that problems/issues only become 'social problems/issues' when someone communicates or makes claims (in public) about them; and b) that the important dimension to study and understand is the *process* through which claims emerge, are publicised, elaborated and contested.

EXERCISE 2.1 Are natural disasters socially constructed?

Using online news sources, take a look at the news coverage of any major natural disaster (e.g. an earthquake, weather-related disasters such as hurricanes, tsunamis or flooding, landslides) in recent times. Examine how the news develops in the first week or so. Are there clearly identifiable phases in the reporting in terms of what topics are focused on, who is interviewed or quoted, how the disaster is framed or interpreted, and so on?

Try and identify some of the key components of the coverage: what does the news coverage focus on – both initially, and after the first few days?

Where are the main sources of information about the disaster? In other words, 'who' defines the nature of the event/disaster for us? Where do they get their information from? What is the balance of 'informed speculation' versus 'first-hand' accounts?

Note how the event/disaster – terrible, obtrusive and highly visual though it is – does not simply convey its own 'meaning'. Rather, the meaning of the disaster is 'constructed' verbally through quotes from experts, victims, rescue personnel, and so on, and through commentary from the media/reporters themselves.

Now take a look at some of the explanations, questions raised and assessments presented in the coverage. At what point does the coverage move from reporting the extent of devastation, loss of life, suffering, and rescue effort, to raising questions about the extent of preparedness for this kind of disaster, including questions not just about emergency planning but also about public investment and policy?

Box 2.1

Are natural disasters socially constructed?

The key argument for the constructionist approach to social problems is the fundamental recognition that problems do not simply exist by themselves in some objective universe, but that they only become 'social' problems when someone draws public attention to them, makes claims in public about them. While this may seem a straightforward explanation for most social issues or problems, the obvious counter-argument in relation to environmental issues is that there are clearly some environmental issues or problems that very much *do* announce themselves by way of their sheer magnitude, visibility and the destruction wrought by them. Thus, earthquakes, hurricanes, tsunamis, flooding, volcanic eruptions, and so on, would all seem to be relatively unpredictable and unforeseeable 'acts of nature', beyond human control and hence beyond any kind of *construction*. Stallings (1990, 1995), Smith (1992, 1996) and others have, however, persuasively shown that even where we are dealing with major natural disasters, we need to call on the constructionist approach to understand the processes by which such disasters come to be defined socially.

Conspicuous, intrusive and devastating as they may be, natural disasters – or for that matter, major unexpected or unforeseen accidents related to man-made structures or processes – do not automatically 'mean' anything, that is, meaning has to be assigned to them or constructed around them. The 'meaning' of devastating floods may be 'divine intervention/retribution', 'nature's revenge' or 'the inevitable results of climate change', but the process of assigning meaning to an event essentially requires the discursive 'work' of claims-makers. Likewise, as Stallings (1995) demonstrates in his analysis of the construction of the 'earthquake threat', the promotion (the choice of word of course is telling) of a natural threat, which is beyond human control, to the status of *social problem* requires much claims-making work through multiple societal forums including scientific (expert panels and committees), political (Congressional hearings), legal and media (national news) forums.

While natural disasters per se are indeed for the most part relatively 'pure' acts of nature, they nevertheless then fit the constructionist mould in terms of – before the disaster strikes – society's policies and preparation for dealing with expected disasters, and – in the aftermath of disaster – the immediately ensuing public arguments about what could/should have been done in terms of social preparedness, how to be better prepared 'next time', how to ensure that new housing estates are not built on flood-plains, or how to ensure that buildings in an earthquake-prone area are built on appropriate foundations and designed to withstand earthquakes.

Objectivity/balance/bias

There is a highly significant and important further dimension to the social constructionist perspective, and one which has particular relevance to media and communications. If social problems are identified as such as a result of processes of claims-making rather than as objective conditions, then the key question for research is not to establish whether a claim is right or wrong, or a true or false representation of a social issue. Rather, the task for research is to establish why and by what means some claims gain prominence and acceptance, while others – which may be equally valid – do not.

This is relevant to media and communication research because it directly counters problematic traditional realist notions of the news media as a 'window on the world' or as a 'mirror representation of reality', and it speaks directly to the classic concerns in news and journalism about accuracy, objectivity, bias/balance and fairness in news reporting. The constructionist perspective enables news research to bypass the futile measurement of accuracy and objectivity in news, futile essentially because one person's accuracy is another person's bias. Concepts such as accuracy, bias and objectivity are not some objective or easily measurable inherent characteristic of news reporting or what is communicated, but depend rather on the perspective/stance/norms/views of those involved as producers or recipients/users and on the context of mediated communication.

A key problem with the core journalistic value of objectivity, particularly where this is translated as being synonymous with giving equal prominence to opposing arguments in a public controversy, is that it may often in itself lead to a distortion or misrepresentation of the balance of opinion on a given subject. Prominent examples include media reporting in the 1980s on scientific opinion about the safety/risk of nuclear power (Rothman & Lichter, 1987) and analyses of media coverage of the climate change debate (Boykoff & Boykoff, 2004). They show that concern with providing objective and 'balanced reporting' results in giving the impression that scientific opinion on the causes and consequences of climate change is split down the middle, when indications from, for example, reviews of scientific publications about climate change (Oreskes, 2004) indicate a near-total consensus among climate scientists. While journalistic notions of objectivity and balance continue to feature strongly, there is increasing recognition not just among media and communication researchers, but among those regulating the media and communications environment and public communications organisations, that rather than there being some objective standard of accuracy or truth by which balance and objectivity can be measured/monitored, it is more productive to ask questions about whose accounts or definitions become prominent or successful in the public sphere, by which means and in whose interest (Hansen, 2016). And these are, of course, questions anchored in the constructionist perspective that issues/problems become defined as such through what is said/communicated about them.

Constructionism and media/communication

If we accept the constructionist argument that environmental problems – and social problems generally – do not 'objectively' announce themselves, but only become recognised as such through the process of public claims-making, then it is also immediately clear that media, communication and discourse have a central role and should be a primary focus for study. In light of this, it is perhaps surprising that the early formulations of the social constructionist perspective offered relatively little comment on media and communications. The development of a social constructionist perspective in mass media and communications research was left to sociologists with a particular interest in communications, notably such prominent American sociologists as Harvey Molotch, Herbert Gans and Gaye Tuchman.

The constructionist argument has implications for understanding media roles, both in relation to how claims are promoted/produced through the public arena of the media and for understanding how the media are a central, possibly *the* central, forum through which we, as audiences and publics, make sense of our environment, society and politics. This boils down to the argument that most of what we as individuals know, we know, not from direct experience (experiential knowledge), but from the symbolic reality constructed for us through what we are told (by friends, family, teachers and other 'officials' of a host of social institutions: schools, government departments or agencies, social institutions, religious institutions, organisations or associations that we belong to, etc.) or read about or hear/see *re*-presented to us through media and communications of various kinds. The centrality of mediated communication in this context is further emphasised by the fact that much of the symbolic construction of reality by a host of social institutions is now itself principally encountered in mediated form rather than through direct contact, observation or engagement.

Public agendas and power

The social constructionist perspective's emphasis on 'claims-making' in public arenas as the constitutive component in the creation of 'social problems' usefully draws our attention to the importance and centrality of getting issues of concern onto the public - and more significantly, the political - agenda.

In this respect, it thus has interesting similarities to the traditions of research in political science and in communication research known as 'agenda-building' and 'agenda-setting', which in turn link with key traditions in the study of 'power' in society. An early and often-quoted formulation from political science which inspired the agenda-setting tradition in media and communication research was Bernard Cohen's statement that 'The press may not be successful much of the time in telling people what to think, but it is stunningly successful in telling its readers what to think *about*' (Cohen, 1963: 13). In other words, the 'power' of the media to influence public and political processes resides principally in signalling what society and the polity should be concerned about and in setting the framework for definition and discussion of such issues.

In their work on agenda-building, political scientists Roger Cobb and Charles Elder (1971) in their discussion of the political process likewise pointed to the centrality of *agendas* recognised by the polity and as forums for the public/political definition of *issues* of conflict 'between two or more identifiable groups over procedural or substantive matters relating to the distribution of positions or resources' (p.32). In contrast to earlier pluralist perspectives on 'power' in society, which had focused on 'decision-making' as a central component of the exercise of power, these perspectives recognised that the ability to control or influence which issues get onto the public/political agenda in the first place was itself a core part of exercising power in society (Lukes, 1974).

The tendency in media research has very predominantly been to focus on the issues that *do* make it onto the media and public agenda – perhaps not least because these are conspicuous and lend themselves most easily to being studied. However, media research has contributed rather less and had significantly less to say about the type of claims-making or publicity management that is aimed principally at keeping issues off or away from the public and/or media agenda. As sociologists critical of power perspectives focused on 'decision-making' were pointing out around the same time as the social constructionist perspective emerged, the ability to keep issues from appearing on the political agenda and thus to ensure that they don't become issues for decision-making, in other words, that they remain 'non-decisions', is as much an exercise of power as making decisions about issues that *are* on the agenda. Edelman (1988) takes this argument a step further by hinting that the placing of certain issues on the public agenda simultaneously achieves the granting of 'immunity' to those issues that are not on the public agenda: 'Perhaps the most powerful influence of news, talk, and writing about problems is the immunity from notice and criticism they grant to damaging conditions that are not on the list' (Edelman, 1988: 14).

While the constructionist perspective then generally focuses our attention on the importance of propelling *claims* into/onto the public arena, it is also clear that an important aspect of the claims-making process may be to keep issues from emerging in particular (public) forums, and thereby influencing the degree to which issues become recognised (or not, as the case may be) as candidates for public concern, discussion or political decision-making.

Issue careers

The emphasis on the *process* of claims-making also led early constructionists to identify the distinctive stages/phases that social problems pass through as they emerge, are elaborated, addressed, contested and perhaps resolved, in other words the 'career' path of social problems. Spector & Kitsuse (1973) thus suggest a four-stage natural history model to describe the career of social issues. Downs (1972), in a much quoted article (not least in studies of environmental, science and risk issues), similarly proposed what he called an 'issue-attention cycle' to explain the cyclical manner in which various social problems suddenly emerge on the public stage, remain there for a time, and 'then – though still largely unresolved – gradually [fade] from the centre of public attention' (Downs, 1972: 38). Downs suggests – and he happens to use 'ecology' or environmental issues as his example, which may in part account for the frequency with which his article has been cited in research on media and the environment – that the career of public issues takes the shape of five distinctive stages: 1) a pre-problem stage; 2) alarmed discovery and euphoric enthusiasm; 3) realising the cost of significant progress and the sacrifices required to solve the problem; 4) gradual decline of intense public interest; and 5) the post-problem stage, where the issue has been replaced at the centre of public concern and 'moves into a prolonged limbo – a twilight realm of lesser attention or spasmodic recurrences of interest' (Downs, 1972).

Criticism of natural history models has focused on the notion that they offer a much too simplistic and linear model of the evolution and progression of social problems. As Schneider (1985) points out, Wiener (1981) for example 'argues that the sequential aspect of natural history models probably misleads us about the definitional process. She believes a more accurate view is one of "overlapping", simultaneous, "continuously ricocheting interaction" (Wiener 1981: 7)' (Schneider, 1985: 225). Wiener's evocative metaphor of 'continuously ricocheting interaction' is a particularly prescient and apt formulation relevant to communication research, as it directly counters the long-dominant – in communication research – linear view of communication and begins to capture (as I shall argue more fully in Chapter 7) the highly dynamic and interactive nature of social communication processes.

Hilgartner and Bosk (1988) similarly criticise natural history models for their 'crude' suggestion of 'an orderly succession of stages' and for inadequate recognition that 'Many problems exist simultaneously in several "stages" of development, and patterns of progression from one stage to the next vary sufficiently to question the claim that a typical career exists' (p.54). While these are valid points, they may not in fact be entirely fair criticisms of what was actually proposed by Spector and Kitsuse and others. Spector and Kitsuse thus never suggested a simple linear progression, but indeed emphasised the heuristic nature of the model and, more importantly, the open-ended nature of the processes described. Downs, particularly with his description of the fifth stage of issue careers and of course through the deliberate use of the word 'cycle', similarly implied a recursive, cyclical process replete with loops.

It is important to note that Downs's issue-attention cycle is not synonymous with a 'media-attention cycle'. Spector and Kitsuse, and indeed Downs, discuss the general social career of issues (part of which may relate to the media) but they never propose that the media career is synonymous with the social career or indeed that – as some communication researchers seem to have assumed – the social career can be 'read off' or deduced from the mapping/charting of the media career.

That the latter is particularly problematic is also clear from numerous agenda-setting studies (discussed more fully in Chapter 7) which have indicated that media coverage is *not*, in Schoenfeld et al.'s (1979) words, a good 'thermometer' or indicator of public sentiment or concern about the environment; for example, a drop in media attention does not necessarily imply that the public has lost interest in an issue or has ceased to be committed to dealing with a social problem.

While heeding then the advice from Wiener, Hilgartner and Bosk and others, that the career of social problems rarely follows a simple linear trajectory, the natural history models remain useful as heuristic models for the simple reason that they focus our attention on the notion of a *career*, the idea that issues evolve over time, and the idea that there are distinctive phases or stages in this process. They further alert us to the notion – and this is perhaps the most significant dimension – that issues don't simply evolve in some vague, general or abstract location called 'society', but rather that they develop and evolve in particular social arenas (Hilgartner and Bosk, 1988) or forums, including political forums and the media, which: a) interact with each other in important ways that determine how issues evolve; b) *host* the stages of problem definition; and c) themselves influence or frame these stages in important ways.

The ups and downs of media coverage of the environment

Downs (1972), in his aptly titled article 'Up and down with ecology – the issue attention cycle', used the case of environmental issues to illustrate the stages in the 'career' of social problems. Numerous studies since have provided ample evidence that media coverage of environmental issues certainly goes up and down in seemingly cyclical patterns, which bear some resemblance to Downs's model, while also confirming that there are many more aspects to what drives this process than can be accounted for within Downs's model.

In very general terms, longitudinal studies of media coverage of environmental issues (Djerf-Pierre, 2013; Schäfer et al., 2014; Hansen, 2015a) show that media interest in environmental issues began in the mid-1960s, increasing to an initial peak in the early 1970s, then declining through the 1970s and early 1980s, followed by another dramatic increase in the latter half of the 1980s, peaking around 1990, then receding again through the 1990s, only to experience a considerable new resurgence in the first decade of the 2000s, followed – particularly in relation to climate change coverage – by further ups and downs. This broad-brush characterisation inevitably obscures the very significant variations from issue to issue as well as the much more frequent ups and downs which can be observed within each of these broad periods. The purpose of painting this highly abstracted general trend is to make three key points about general media coverage of environmental issues.

The first point is that the concept of 'environmental issues' or of the 'environment' as a social problem only emerged on the public agenda in the 1960s. This is not to say that the media had not covered such issues as pollution or man-made erosion before, but instead that the new perspective or 'framework' of ecology and its associated more holistic view of the environment only emerged on the public arena in the 1960s. Longitudinal studies have also confirmed that the 'environmental/ecological' perspective which emerged in the 1960s gradually but steadily consolidated in media coverage over the next few decades, to the extent that Einsiedel and Coughlan (1993: 141) were able to conclude that the 'environment' in environmental reporting had become 'vested with a more global character, encompassing attributes that include holism and interdependence, and the finiteness of resources'.

The second point of this broad generalisation is to emphasise that society seems to go through broad phases of varying receptiveness to issues such as 'the environment'. When trying to map and explain the more detailed level of media coverage (and newsworthiness) of specific issues, media researchers need to take into consideration the wider 'climate of opinion' characteristic of the particular historical period under scrutiny. In other words, it is instructive to consider whether we are, broadly speaking, within a generally 'receptive' period or whether we are going through a period where social and political sentiment is driven by concerns about economic and industrial decline seen as associated with globalisation and relinquishing control over movement of goods and labour, and, in turn, a perception that 'environmental protection' is an obstacle to economic growth.

EXERCISE 2.2 Interacting issue-agendas

The late 1980s witnessed a considerable surge in social, political and media interest in the environment. Media analyses have shown how the amount of coverage of environmental issues increased dramatically in the latter half of the 1980s towards a peak in the very early 1990s. Then, rather abruptly, the intensity of interest dropped away again during the first half of the 1990s. Numerous factors, as we shall see in more detail in the following chapters, influence these 'ups' and 'downs', but it has been suggested that the global economic downturn of the early 1990s contributed significantly to the decline in media – and perhaps public – interest in the environment and environmental protection. Bluntly put, tough economic pressures meant that people had other more pressing things to worry about than 'saving the environment'.

However, this is likely a much too simplistic explanation. Consider the global economic downturn and crisis that started towards the end of the first decade of this century, and note how political rhetoric about saving or restoring the economy was not automatically linked to calls for less environmental regulation. Consider how and the extent to which political and public discourse about 'rescuing the economy' has gone hand in hand with discourse on how to deal with climate change and its likely effects. To what extent have economic crisis and environmental crisis been construed as part of the same problem – demanding concerted action and solutions?

The third and final purpose is to make the point that what we might, equally broadly, call the environmental/ecological paradigm, once introduced in the 1960s, has remained firmly on the media and public agenda ever since, and while it's had distinctly more wind in the sails during some periods than others, and while its history has been characterised by all the conventional hallmarks of cycles of claims and counterclaims, we have not as yet seen any hints that a wholesale paradigm shift is likely to occur any time soon. In fact, it could be argued that the environmental paradigm – far from showing signs of increasing fragmentation – has become increasingly holistic through the rise of global concerns about climate change.

The notion of 'threshold events' or 'tipping points' is also key to understanding what drives the dynamics of media and public communication about the environment. Thus, despite warnings from scientists, ecologists, energy experts and others about the depletion of energy reserves, it took the threshold event of the Arab oil embargo of 1973–74 to propel this issue into a position of media prominence (Schoenfeld et al., 1979; Mazur, 1984). Mazur (1984) likewise demonstrates how serious nuclear power plant accidents in the 1970s went entirely unreported, while some minor (by industry and scientific standards) incidents which occurred

Box 2.2

Issue-interaction, claims-making and threshold-events/tipping points

The social constructionist perspective helpfully draws our attention to the central and all-important role of 'claims-making', that is, that issues only become issues for public and political concern if someone draws attention to them and makes claims about them. A cursory look at the ups and downs of media coverage of environmental issues however also tells us that claims-making may not in and of itself be sufficient to ensure widespread publicity and media coverage for an issue. Crucially, claims need to (be made to) resonate with wider public interests/concerns/fears to be successful. As Ungar (1992: 484) puts it:

> Recognition in public arenas, which is a *sine qua non* of successful social problems, cannot be reduced to claims-making activities, but depends on a conjunction of these *and* audience receptiveness. Claims-making, after all, can fall on deaf ears or meet with bad timing.

Longitudinal studies have shown that major environmental disasters – rather than detracting attention from other environmental issues – tend to have a focusing or galvanising effect on the overall prominence of environmental issues in media and public communication. Longitudinal studies of news coverage of the environment thus point to synergistic interaction of different categories of environmental issue, showing that 'when one issue category receives a high level of attention, the attention to other environmental issues tends to increase as well' (Djerf-Pierre, 2012a: 299).

subsequent to the Three Mile Island accident in 1979 received considerable media coverage. A major threshold event such as the Three Mile Island accident thus greatly sensitises the media to similar and related events, and helps increase the sheer volume of coverage and attention devoted to – in this case – nuclear events and issues. It also, perhaps more importantly, helps establish a dominant frame or perspective for subsequent coverage of similar and related issues.

More recently, the most celebrated example of an issue that only gained prominence in the media when a range of important factors began to coalesce is global warming/climate change. In two early analyses, Ungar (1992) and Mazur & Lee (1993) show how sustained claims-making by scientists about the damage to the ozone layer and indications of a process of increased global warming took place for a considerable time without these issues receiving much media attention. This changed dramatically when the seemingly exceptionally dry summer of 1988 (in Northern America and Northern Europe) furnished the media – and public concern – with a direct and immediate reference point. It was not that climatologists and other scientists believed that the dry summer of 1988 was anything other than a 'normal' occurrence in cyclical weather patterns, but it provided a fertile context for the promotion of claims about global warming/climate change as caused by harmful human practices.

The apparent cyclical trends and the apparent poor fit between amount of media coverage and claims-making (e.g. by scientists), which studies have found in analyses of media coverage of selected environmental issues thus tell us: 1) that sustained claims-making about an environmental issue may not in itself be sufficient to secure its prominence on the media agenda; and 2) the ups and downs of issues on the media agenda cannot be taken as evidence of the seriousness of the issue or of whether the issue has been resolved or appropriately addressed through legislation, allocation of resources, research, and so on.

Trumbo (1996), McComas & Shanahan (1999), Mikami et al. (2002) and Brossard et al. (2004) have all usefully drawn on Downs's 'issue-attention' cycle for explaining the cyclical phases of media coverage of global warming/climate change, as have Nisbet and Huge (2006) in their comprehensive study of a different environmental issue, plant biotechnology.

Trumbo's (1996) analysis of climate change coverage over the decade 1985–1995 provides a particularly clear early application of Downs's framework. Trumbo shows that the coverage 'fits Downs's five-stage model fairly well' (1996: 276), and, for example, confirms the prediction that while issues rise and fade in prominence, once introduced on the media agenda, attention – even when their prominence fades in the 'post-problem stage' – remains above the low levels seen before the initial increase in attention (the 'pre-problem stage').

The most comprehensive and systematic evidence to date on issue-attention cycles in media coverage of environmental issues comes from the studies by Monica Djerf-Pierre (2012a, 2012b, 2013) in her longitudinal analysis of Swedish television news coverage of the environment from 1961–2010, and by Mike Schäfer et al. (2014) in their analysis of climate change coverage in Australian, German and Indian print media from 1996 to 2010.

Schäfer et al. (2014) confirm that – country and media-system differences notwithstanding – the coverage of climate change follows remarkably similar trends, with clearly identifiable ups and downs, but also with an overall significant increase in media attention over the 15-year period examined. They are further able to demonstrate the relative contribution of selected factors/drivers on the prominence of climate change coverage in the newspapers. This analysis shows that contrary to what has been suggested by many previous studies, "weather and climate characteristics are not important drivers of issue attention". Neither do they find evidence that scientific publications impact to any significant degree on the amount of media attention. By contrast, it is clear from their analysis that domestic political attention, NGO activity and international summit events are major drivers of media attention.

Djerf-Pierre's study of Swedish television news coverage of environmental issues over the half century from 1961 to 2010 is valuable for its comprehensive demonstration that despite multiple cycles of ups and downs, these become less frequent over time and the overall trend is up, and she is able to conclude that her analysis thus provides clear evidence of the "long-term institutionalization of the environmental domain in television news journalism". The longitudinal nature of the study and the differentiation between different environmental issues/themes in the coverage further enable Djerf-Pierre to advance Downs's issue-attention framework beyond the cycles of single issues, and to show how larger clusters of issues form 'meta-cycles' – with a rather larger time-interval than single issues – of peaks and troughs in media attention to environmental concerns.

While comparisons of studies of media coverage of climate change in different countries give a clear indication that a range of factors (including media organisational arrangements, political 'leanings' of particular media organisations, journalistic practices and values, and perhaps wider culturally determined agendas – discussed in more detail in the next two chapters) impinge on the precise ups and downs of media coverage, such comparisons have also provided tantalising evidence of broader cyclical patterns that resonate well with stages in Downs's issue-attention cycle. Where Downs's model is particularly helpful is in providing a framework for identifying and making sense of the different stages in the cycle of media coverage – that is, it enables us, when looking at the ups and downs of media coverage, to begin to answer the question, 'What drives the coverage at this particular stage, and what is going on during this particular period of coverage?'. It is of course crucial to keep in mind that Downs's model concerns the general social career of issues – it was not designed specifically for explaining the *media* career of an issue, and we should not lose sight of the fact that the media are just one of the public arenas in which social problems are articulated.

Claims-making and framing

The key achievement of the constructionist perspective on social problems lies in the recognition that problems do not become recognised or defined by society as problems by some simple objective existence, but only when someone makes claims in public about them. The construction of a problem as a 'social problem' is then largely a rhetorical or discursive achievement, the enactment of which is perpetrated by claims-makers, takes place in certain settings or public arenas and proceeds through a number of phases.

American sociologist Joel Best (1995, 2013) suggests that analysis of the construction of social problems needs to focus on: 1) the claims themselves; 2) the claims-makers; and 3) the claims-making process. To these we might add the public arenas (Hilgartner & Bosk, 1988) or settings (Ibarra & Kitsuse, 1993), as arenas or settings (including the media, the courts, parliamentary politics, the scientific community) set their own boundaries for or impose their own constraints on what can and cannot be said. I shall discuss the constraints of the media in more detail in the next chapter, as well as the actors involved in the claims-making process. Here, however, I will focus on the rhetorical/discursive aspect of the claims-making process by drawing on Ibarra and Kitsuse's (1993) notion of 'vernacular resources' and on the concept of framing.

In an exceptionally clear exposition, Ibarra & Kitsuse (1993) restate the emphasis in Spector & Kitsuse's original statement about social problems as claims-making, that language or discourse is at the heart of the construction of social problems ('As parts of a classification system, condition-categories [referred to in the original statement as 'putative conditions'] are first and foremost units of language', 1993: 30). The object of study is therefore, as Ibarra and Kitsuse put it – borrowing from Mills (1940) – the 'vernacular displays' (p.29) of those involved in the claims-making process. They proceed to outline the core foci of constructionist analysis under the following five headings:

- '*Rhetorical idioms* are definitional complexes, utilizing language that situates condition-categories in moral universes. (…) Each rhetorical idiom calls forth or draws upon a cluster of images' (34). Examples:
  - The rhetoric of loss
  - The rhetoric of unreason
  - The rhetoric of calamity.
- The *counter-rhetorics* are discursive strategies for countering characterisations made by claimants. They tend to be less synoptic or thematic: For example, instead of arguing for ozone layer destruction, it is the claimant's description, proposed remedies or something other than the candidate problem that is rebutted. These counter-rhetorics tend not to counter the 'values' conveyed in the rhetorical idioms so much as they address their current application and relevance (1993: 34–35). Counter-rhetorics divide into:
  - Sympathetic counter-rhetorics, including such rhetorical strategies as naturalising, costs involved, declaring impotence, perspectivising, and tactical criticism
  - Unsympathetic counter-rhetorics, including such rhetorical strategies as antipatterning, the telling anecdote, and the counter-rhetorics of insincerity and hysteria.
- *Motifs* are recurrent thematic elements, metaphors and figures of speech that encapsulate, highlight or offer a shorthand to some aspect of a social problem. Examples: *epidemic, menace, scourge, crisis, blight, casualties, tip of the iceberg, the war on* (drugs, poverty, crime, gangs, etc.), *abuse, hidden costs, scandal, ticking time bomb* (47). Ibarra and Kitsuse point to the particularly pertinent 'need for understanding their *symbolic currency,* that is, why some motifs are prized while others are considered best avoided' (48).

- *Claims-making styles* shift attention from the language in which claims are cast to the bearing and tone with which the claims are made. For example, claimants (and counterclaimants) may deliver their claims in *legalistic* fashion or *comic* fashion, in a *scientific* way or a *theatrical* way, in a *journalistic* ('objective') manner or an 'involved citizen' (or '*civic*') manner (1993: 35).
- *Setting*: How do the formal qualities of particular settings structure the ways in which claims can be formulated, delivered and received? What kinds of rhetorical forms can be employed because of the imperatives or conventional features constituting the various locations? What are the various categories of persons populating these settings, and how do their characteristics entail accountably interacting with claims and claims-makers? (pp.53–54).

The last question is particularly relevant to media and communications researchers as it points directly to core concerns about media organisational arrangements and journalistic practices/values, which structure the relationship with sources (claims-makers) and impact on the articulation and framing of claims.

There is considerable overlap and compatibility between these rhetorical dimensions and the framework proposed by Gamson (1988), who talks about 'packages' and refers to many of the same linguistic features and rhetorical devices as Ibarra and Kitsuse. Where Ibarra and Kitsuse's scheme is particularly useful is in its inclusion of 'claims-making styles' and 'settings' as key factors influencing what can be said, how and with what implications.

Although Ibarra and Kitsuse (1993) do not explicitly position themselves within the context of 'framing', much of what they say resonates very well with the concept of framing, as it has grown to be used in media and communication research. Reese (2001) for example defines frames as '*organizing principles* that are socially *shared* and *persistent* over time, that work *symbolically* to meaningfully *structure* the social world', while Gitlin (1980) sees frames as 'principles of selection, emphasis, and presentation composed of little tacit theories about what exists, what happens, and what matters'. Frames in other words draw attention – like a frame around a painting or photograph – to particular dimensions or perspectives, and they set the boundaries for how we should interpret or perceive what is presented to us (i.e. is the glass described as 'half full' or 'half empty'?). Gamson (1985), with particular reference to the media, suggests that 'news frames are almost entirely implicit and taken for granted. (…) News frames make the world look natural. They determine what is selected, what is excluded, what is emphasised. In short, news presents a packaged world' (Gamson, 1985: 618).

Miller and Riechert (2000: 46) make an important further addition to the definition of framing by suggesting whose interests are served. They thus suggest that framing is usually thought of as 'driven by unifying ideologies that shape all content on a topic into a specific, dominant interpretation consistent with the interests of social elites'. While this recovers an important Marxian/Gramscian element by suggesting that frames generally work in the interest of powerful classes or elites in society, we should also recognise that framing can, in principle, be made to work for any social group or interest. However, the framing task is clearly much more challenging and difficult for those who are working 'against the grain' by trying to un-seat a dominant, culturally deep-seated, interpretive package or frame, than for those who are able to anchor their arguments firmly within a dominant interpretative framework and thus able to work 'with the flow' rather than against it.

In a synoptic – and much quoted – overview of the 'framing' concept and its various disciplinary origins, Entman (1993) defines framing as follows:

> To frame is to *select some aspects of a perceived reality and make them more salient in a communicating text, in such a way as to promote a particular problem definition, causal interpretation, moral evaluation, and/or treatment recommendation* for the item described.
>
> (52; emphasis in original)

Claims-makers, including media and news professionals, then draw attention to particular interpretations through *selection* (i.e. our attention is drawn to some aspects while others, not selected, are kept out of view) and *salience or emphasis*, which promotes particular definitions/interpretations/understandings rather than others. Perhaps the most significant point about how a problem is defined, is that the definition invariably carries with it the allocation of responsibility or blame as well as – implicitly or explicitly – directions for the problem's solution. As Entman argues:

> All four of these framing functions [problem definition, causal interpretation, moral evaluation, treatment recommendation] hold

> together in a kind of cultural logic, serving each other, with the connections cemented more by custom and convention than by the principles of valid reasoning or syllogistic logic. The two most important of these functions are the problem definition, since defining the problem often virtually predetermines the rest of the frame, and the remedy, because it promotes support of (or opposition to) actual government action.
>
> (2003: 417–418)

This is succinctly captured by Charlotte Ryan (1991: 57), who argues that the notion of framing directs the analysis of claims-making and the construction of social problems to ask three core questions: 1) What is the issue? (definition); 2) Who is responsible? (identification of actors/stakeholders); and 3) What is the solution? (suggested action/remedies). And to answer these questions we can usefully draw on the analytical framework offered by Gamson and Modigliani (1989) which helpfully sets out the notion of a 'signature matrix' to indicate the constituent parts of frames – in other words, enables us to answer the question, by which rhetorical/linguistic and other devices is a frame constituted and sustained?

Gamson and Modigliani's (1989) discussion of framing is particularly useful because it draws attention to the two core meanings of 'framing' in media research: on the one hand, framing as a term for the stories/discourses/ideologies/packages available to us for making sense of our environment, and, on the other hand, framing as the workings or operation of the constituent parts that together contribute to a particular frame. Gamson and Modigliani thus suggest 'that media discourse can be conceived of as a set of interpretive packages that give meaning to an issue. A package has an internal structure. At its core is a central organizing idea, or frame, for making sense of relevant events, suggesting what is at issue', and that 'a package offers a number of different condensing symbols that suggest the core frame and positions in shorthand, making it possible to display the package as a whole with a deft metaphor, catchphrase, or other symbolic devices' (p.3).

The constituent parts, which together contribute to or build the 'frame', can, according to Gamson and Modigliani (1989), be identified as five *framing devices* that suggest how to think about the issue (metaphors; exemplars [i.e. historical examples from which lessons are drawn]; catchphrases; depictions; visual images [e.g. icons]) and three *reasoning devices* that justify what should be done about it (roots [i.e. a causal analysis]; consequences [i.e. a particular type of effect]; and appeals to principle [i.e. a set of moral claims]).

What is offered, then, is an analytical framework for characterising and un-packing the interpretive packages in 'a signature matrix that states the frame, the range of positions, and the eight different types of signature elements that suggest this core in a condensed manner' (Gamson & Modigliani, 1989: 4).

Both Ibarra and Kitsuse's discussion and, perhaps more so, Gamson and Modigliani's analytical model should then, I suggest, be seen as essentially suggestive lists of key questions/tools for identifying and analysing the devices and frames that come into play in the claims-making process and for identifying and analysing the rhetorical means by which some claims-making is more successful than other types of claims-making.

EXERCISE 2.3 Claims-makers, rhetorical devices, settings and frames

Using any major online news provider, identify a recent news item about any environmental issue.

Why is this story in the news on this particular day? What or whose action, in what forum or setting, has caused the topic/subject of this story to be 'news'?

Who are the key claims-makers quoted or referred to in the story? Are they scientists, experts, politicians, business/pressure group/agency representatives, or 'ordinary' people?

What kind of forum or 'setting' do they represent? How – if at all – do the conventions of the forum/setting seem to impinge on or shape what is being said about the topic/subject of the news story?

Are there any words/terms in the report that 'stand out' either as characteristic of a particular discourse (e.g. a science discourse or a legal discourse), or as examples of what Ibarra and Kitsuse refer to as motifs, e.g. 'epidemic', 'war on ...', 'tip of the iceberg', 'ticking time bomb', etc.?

How does the way that the news is constructed, including the particular terms/words used, shape the answer to the three framing questions: 1) what is the issue/problem? 2) who/what is to blame? and 3) what is the (implied) solution?

Conclusion

The constructionist perspective provides a framework for analysing and understanding why some environmental issues come to be recognised as issues for public and political concern, while others – potentially equally important issues – never make it into the public eye, and thus fail to command the political attention and resources required for their resolution. The constructionist perspective focuses our attention on the role of claims-makers and on the public definition of social problems as essentially a rhetorical/discursive achievement. Drawing on the notion of 'vernacular resources' and on the concept of framing, the chapter outlined a number of key analytical foci and tools for examining the construction and contestation of environmental issues: these included Ibarra and Kitsuse's rhetorical idioms, counter-rhetorics, motifs, claims-making styles and settings; Gamson & Modigliani's focus on media or meaning 'packages' and their constitution through a 'signature matrix' of framing and reasoning devices; and the concept of framing understood as selection and salience, communicating a particular problem definition, which in turn carries with it implied causes, moral evaluations and solutions.

The constructionist perspective further shows that social problems do not simply appear in some vague location called society, but that they are actively constructed, defined and contested in identifiable public arenas – notably the media – and that the careers of social problems are characterised by distinctive stages or phases. While claims-making and definition takes place in a number of arenas, the media are a particularly important arena or hub, because it is through the media that we as publics predominantly learn about what goes on in other key arenas (such as parliament, science or the courts). But the media are not simply an open stage; as a public arena they are governed by their own organisational and professional constraints and practices, some of which have been signalled in this chapter (e.g. the journalistic value of 'objectivity' and 'balance'; the role of 'trigger-events'), while others will be discussed in more detail in the following two chapters.

Further reading

Schneider, J. W. (1985). Social problems theory: the constructionist view. *Annual Review of Sociology, 11*, 209–229.

An early overview of what was then still an emerging perspective; particularly good referencing of the key roles of media and communication in the construction of social problems.

Best, J. (2013). Constructionist social problems theory. In: C. T. Salmon (Ed.), *Communication Yearbook 36* (Vol. 36, pp. 236–269). London: Routledge.

A succinct overview and updating of constructionist social problems theory by one of the leading constructionist sociologists of our time.

Hannigan, J. A. (2014). *Environmental Sociology* (3rd ed.). London: Routledge.

Now in its third edition, Hannigan's book continues to offer one of the clearest introductions to constructionist environmental sociology – see particularly *Chapter 3: Social construction of environmental issues and problems.*

3 Making claims and managing news about the environment

THIS CHAPTER:

- Applies the previous chapter's introduction to the constructionist perspective to news showing that there is little that is 'natural' about environmental news; even 'natural disaster news' can best be understood in terms of something which has to be actively constructed.
- While recognising the importance of news values and of the political, economic and cultural factors that circumscribe the production of news, this chapter focuses on the activities and communication strategies of claims-makers.
- Examines claims-makers in the broadest sense of the term, including environmental pressure groups, government and business/industry.
- Discusses the distinction between 'insider' and 'outsider' groups and the extent to which different types of claims-maker seek or depend on media coverage.
- Introduces the key claims-making tasks as: 1) commanding attention; 2) claiming legitimacy; and 3) invoking action.
- Examines how claims-makers (environmental pressure groups as well as corporate business) have sought to exploit the Internet and other newer forms of communication for information, news management and campaigning purposes.
- Discusses corporate image management strategies.

Introduction: the 'constructed-ness' of news

Considering the quantity of news coverage of natural events, disasters and emergencies associated, as one is told, with climate change and related environmental damage during much of the first two decades of the 21st century, one could be forgiven for assuming that the environment is by nature and by definition 'newsworthy'. Even (contrary to what was argued in the previous chapter) that the environment becomes news almost by itself, and that there is little or nothing 'constructed' about either the prominence of environmental news or the nature of environmental topics that are communicated about in the public sphere. This would perhaps particularly seem to be the case with major natural disasters as well as with smaller-scale regional emergencies, such as recurrent flooding and associated damage.

Major natural disasters from this century that stand out as singularly 'newsworthy' in and of themselves include: one of the deadliest natural disasters in history, the Indian Ocean earthquake of 26 December 2004, which triggered a series of tsunamis wreaking havoc and devastation in countries bordering the Indian Ocean – most catastrophically Indonesia, Thailand, India and Sri Lanka; Hurricane Katrina in August 2005; the earthquake, one of the most powerful in recorded history, off the coast of Japan in March 2011 and the subsequent tsunami and devastation, killing more than 18,000 people; Hurricane Sandy 2012 causing widespread devastation in the Caribbean and the east coast of the United States; and in August 2017, the most extensive monsoon-related flooding in years in Bangladesh, India and Nepal, killing over a thousand and leaving millions homeless (Farand, 2017).

It only takes, however, a slightly closer – and a slightly longer-term – look at such coverage to realise that there is little or nothing that is 'natural' about news coverage of the environment or environmental events/disasters. The first anomalies begin to appear when observing how some major disasters, having received masses of news coverage for a while, then suddenly all but disappear from the media agenda, not because they have been alleviated or resolved, but perhaps because other events, disasters or issues have literally pushed them off the news agenda in what Hilgartner and Bosk (1988) have aptly described as the intense competition for space on a media agenda with limited 'carrying capacity'. Occasionally, the competition for news space is even between disaster stories of a similar type. For example massive coverage of the Hurricane Harvey destruction in Houston, Texas and elsewhere in the same region of the USA in August 2017 to some extent squeezed out media attention to the ongoing problems in the wake of a devastating landslide in Sierra Leone (e.g. Boyle, 2017) or the monsoon-related flooding and devastation in Bangladesh, India and Nepal in the same month. But a comparison of the nature and extent of news coverage of such events must also take into consideration the very different geopolitical 'news status' of countries or regions involved, as well as practical factors to do with news and communications access (proximity to major news centres, location of reporters, and – increasingly – the availability of mobile phone technology amongst citizens affected and able to provide first-hand accounts).

Another key indication of the complex and 'un-natural' processes behind news coverage emerges when our attention is suddenly drawn to major environmental disasters or problems which have existed and been developing for a considerable length of time, but only now show up on the news radar: numerous instances of coverage of drought, famine and related sociopolitical upheaval in sub-Saharan Africa over several decades fall into this category; so too do some instances of flood devastation, which, although seemingly sudden and unexpected, may have been caused or exacerbated by years of human damage (e.g. logging) to natural flood defences. Wide variations in both the amount and nature of news coverage of environmental disasters/emergencies depending on *where* geographically they happen and *who* is affected further alert us to the continued significance of Galtung and Ruge's (1965) classic study of news values, particularly the importance of geopolitical and 'cultural proximity' as a key determinant of what gets covered, how and for how long.

If further indication were needed of the complex processes governing news coverage of the environment, consider the evidence from numerous studies of long-term trends in environmental news coverage. Invariably, such studies have shown what Downs in his prescient article published in 1972 aptly referred to as the 'ups' and 'downs' of coverage: the quantity and nature of coverage given to the environment and environmental issues has gone up and down in phases that bear little relation to the (scientific) discovery, material existence (as, for example, measured through scientific, economic or social indicators), persistence or resolution of environmental problems.

In order to understand the ups and downs in media coverage, as well as to understand why some issues receive much more coverage than others, we need, then, to look closer at the complex processes which influence the production of news. Environmental news, in other words, is not just something that happens and automatically gets recorded and reported by the media. News, as numerous key news studies have amply demonstrated, is made, created and selectively reported. The making of news is a complex process of interaction between, on the one hand, institutions and individuals in society who act as sources or subjects of news, and on the other hand, the news media whose own organisation and professional practices influence what institutions, events and individuals get reported. Further layers of influence circumscribing the production of news include *political*, *economic, technological* and – not least (see also Chapter 4) – *cultural* factors.

In this chapter, we focus on media-external influences on the communication process, on the communication strategies and influence of key claims-makers and stakeholders in environmental debate and controversy. The type of claims-maker that perhaps most easily comes to mind in relation to environmental debate and controversy is probably environmental pressure groups, but there are clearly many others – government departments, scientists, economists and not least the businesses and industries whose operational practices are often the subject of environmental debate and controversy. Indeed, there is a growing body of research indicating that news sources and claims-makers of all kinds have become increasingly focused on influencing and managing – through what is commonly now referred to as 'strategic communication' (see Heath and Johansen, 2018) – the public communications environment both in relation to traditional print and broadcast news media and in relation to online and social media.

Claims-making, social construction and key tasks

If, as argued in the previous chapter, the construction of the environment as a social problem depends on successful public claims-making, then the media constitute a key public arena, in which the voices, definitions and claims of claims-makers (notably representatives of government, public authorities, formal political institutions, professional communities and associations, pressure groups, etc.) are put on public display and compete with each other for legitimacy. But the media are not merely a convenient public arena or window; they play – through the organisational and professional arrangements of news making – an active role in the construction, inflection and framing of both issues and claims-makers. Gaining media coverage may often seem the most immediate task for claims-makers, but as Ryan states in her book on media strategies for grassroots: 'gaining attention alone is not what a social movement wants; the real battle is over whose interpretation, whose framing of reality, gets the floor' (Ryan, 1991: 53).

'Framing' in media coverage hinges on two dimensions: 1) selection/ accessing of sources/claims-makers; and 2) presentation/evaluation of arguments/actors:

> Framing essentially involves selection and salience. To frame is to *select some aspects of a perceived reality and make them more salient in a communicating text, in such a way as to promote a particular problem definition, causal interpretation, moral evaluation and/ or treatment recommendation* for the item described.
>
> (Entman 1993: 52; emphasis in original)

It is reasonable to expect a fair degree of 'fit' between *who* is quoted in media coverage and *how* issues are framed or defined. Indeed, in a study of US news coverage of climate change, Trumbo (1996), for example, found a strong association between the types of claims-makers (scientists, politicians and interest groups) accessed by the media and the types and prominence of different frames, leading him to conclude that 'changes occurring in the life-course of this issue [climate change] apparently involved shifts linked to who was getting their message into the media rather than how the media was choosing to present the information' (Trumbo 1996: 281).

But, while analysis may confirm an association between the types of claims-makers accessed and the relative prominence of general thematic clusters or frames in media coverage, such an analysis does not go far enough. Crucially, it does not show us how different sources and their testimonies or claims are framed by the media, or how, through the framing and definitional work engaged in by the media themselves, sources and their claims may achieve legitimacy and credibility through media coverage, or alternatively, may be systematically undermined by such coverage.

In a seminal article, published in 1976, Solesbury makes the crucial distinction (often overlooked by media analysts, pressure groups and others keen to equate amount of media coverage with successful claims-making) between, on the one hand, gaining media coverage for a particular cause or issue, and on the other, ensuring that the tone or framing of the coverage is such that the 'message' conveyed to the larger public is one of legitimacy and support for the claims being made. To these two tasks, Solesbury adds a further crucial task or stage in the claims-making process, namely that of 'invoking action'. The three key tasks for claims-makers making claims about a putative problem, then, are: 1) commanding attention; 2) claiming legitimacy; and 3) invoking action.

Resonant of Downs's (1972) issue-career stages, Solesbury's three stages critically disentangle these three separate achievements (often ignored or confused in the hyped-up rhetoric of claims-makers themselves and indeed in some media analyses) to emphasise that media coverage cannot be assumed to be 'the right kind of media coverage'. Coverage may be positive and be conferring legitimacy, lending credibility to the claims being made, *or* it may be critical, negative, undermining the legitimacy of both claims and claims-makers, or worse still, marginalising and branding the claims-makers as insincere or 'extremist'. Nor does media coverage itself guarantee social or political action, that is, that something is actually done to address or resolve the problem about which claims are being made. As well as providing a sobering yardstick for claims-makers themselves for the planning and evaluation of claims-making strategies, Solesbury's three-tasks list serves as a useful reminder and tool for media-analysts seeking to assess media and communication roles in claims-making and campaigning processes.

Box 3.1

Commanding attention and claiming legitimacy (or perhaps not)

The much studied – and celebrated – case of Greenpeace's action against the Anglo-Dutch oil company Shell over its plans to dump a redundant North Sea oil storage installation, the Brent Spar, in the Atlantic Ocean in 1995 provides perhaps one of the clearest illustrations available of the complex mix of actors/agents, claims, claims-making processes, issue careers, news processes, and so on, which combine to produce a social/political outcome or change. The full findings of several comprehensive analyses of the Brent Spar controversy (Hansen, 2000; de Jong, 2005; Bakir, 2006) will not be reiterated here. Instead, I shall focus on one particular aspect: the difficulties for claims-makers in managing or even influencing the media-framing of their claims and actions. The following is an excerpt from an analysis (Hansen, 2000: 62–66) of the reporting in British newspapers of the 1995 controversy between Greenpeace and Shell over the proposed deep-sea dumping of the Brent Spar.

Framing Greenpeace

Greenpeace and its campaigners were generally referred to as 'Greenpeace', 'protesters', 'campaigners' and 'activists'. However, the *Telegraph*, in keeping with the anti-Greenpeace tone already indicated, and consistent with what linguists have described as the phenomenon of overlexicalisation (Fairclough, 1989), deployed a rather wider and clearly more negative set of descriptors.

Greenpeace and its actions were described variously in terms such as 'nuisance', the formulaic phrase 'single-issue + politics/campaigners/group/pressure-group', 'self-righteous', 'rebels', 'bearded', 'doleful', 'extreme eco-warriors', 'eco-sentimentalists', 'emotional', 'misguided', 'militant group', 'undemocratic', 'irresponsible', 'propagandists', 'arrogant', 'more fervent than competent' (*Daily Telegraph* 7 September), 'bullyboys'. A Greenpeace spokesman 'gibbers', rather than speaks, in a *Sunday Telegraph* quote (25 June). Where the *Mail* and the *Mirror* tended to portray the battle between Greenpeace and Shell in terms of a battle between daring, heroic, homely, idealistic protesters and a 'huge', 'large', 'multinational', greedy 'giant', the *Telegraph*, particularly conscious of the potential representation of the battle over Brent Spar as a David (Greenpeace) versus Goliath (Shell) battle, endeavoured to expose this as a fallacious and mythical representation by focusing repeatedly on the size, (business) value and power of Greenpeace as itself a multinational organisation.

The framing of Greenpeace as a powerful, threatening (to business, democracy and the public) organisation was further emphasised by the metaphoric labelling of its politics as 'environmental jihad' (*Daily Telegraph* 24 June), 'harassment' (*Daily Telegraph* 21 June), a 'black art (…) seducing or bullying public opinion by media manipulation ranging from hype and distortion to public demonstrations and criminal disobedience' (*Daily Telegraph* 5 October).

In contrast to the *Telegraph*'s negative framing of Greenpeace, the *Mail*, although by no means uncritical or uniformly positive, generally framed Greenpeace protesters in a way which showed them as devoted, daring (only the *Mail* and the *Mirror* ever referred to Greenpeace's occupation of the Brent Spar as 'daring'), committed and ingenious.

The portrayal of Greenpeace as the potentially 'threatened', 'peaceful demonstrators', rather than a threatening powerful organisation was further underlined by the *Mail*'s revelation in several articles that the Government had authorised Royal Marine Commandos to be on stand-by to help Shell evict the Greenpeace protesters from the Brent Spar platform.

The incompatibility of this particular news angle with the *Telegraph*'s framing of Greenpeace as a powerful and manipulative organisation may help explain why it received no mention in the *Telegraph*.

In sharp contrast to the wide range of negative terms/labels used by the *Telegraph*, glorifying battle terms such as 'hero/heroes/heroic', 'army of green warriors', 'Greenpeace commandos', 'fighters' and 'daredevils' were unique to the *Mirror*. The *Mirror* from the outset painted the controversy as a fight between the brave and heroic Greenpeace ('HEROES DID GREAT JOB: FISHERMEN PAY TRIBUTE TO GREENPEACE OVER BRENT SPAR', *Daily Mirror* 23 June 1995; 'OIL RIG HERO GETS SHELL DISCOUNT CARD', *Daily Mirror* 24 June 1995; 'DEMO MAN WANTS HUG FROM LOVE: BRENT SPAR HERO AL BAKER ARRIVES BACK IN THE SHETLANDS', *Daily Mirror* 24 June) fighting on behalf of us (the readers) and nature against a greedy, powerful, inflexible and irresponsible multinational company.

The *Mirror* went further than simply legitimating Greenpeace's actions. In a manner typical of the particular mode of readership address often adopted by tabloid papers (see Hall et al., 1978) the *Mirror* cast itself in the role of representing, speaking on behalf of and campaigning on behalf of its readers. The *Mirror* thus not only claimed to be the first to have reported on the scandal and problems of dumping, but also portrayed itself as having been largely instrumental in bringing about the Shell U-turn on dumping of the Brent Spar.

Unlike the other papers, the *Mirror* rhetorically constructed an active role for its readers, a sense of participation, by characterising Shell's change of mind as a result of 'people power' (*Daily Mirror* 22 June and 27 June), by inscribing its readers into the same general 'battle' language used for describing the conflict between Shell and Greenpeace by referencing its readers with the active battle metaphor 'the Mirror's army of readers', by constructing relationships between Greenpeace and its readers (Greenpeace was quoted as thanking the *Mirror*'s readers for their support; *Daily Mirror* 22 June), and by addressing its readers directly using the personal pronoun 'you' (e.g. '£700M: YOU SHELL OUT TO CLEAN UP 200 RIGS: YOU'LL PAY £700M FOR OIL FIASCO' *Daily Mirror* 22 June).

Public arenas and source-power

While many sources and voices contribute to the claims-making process which helps give some environmental issues visibility and prominence in public arenas like the news media, it is also clear from a long tradition of studies in the sociology of news, that access to the media is highly selective, unequal and hierarchical. In an early formulation, Hall (1975) thus argued, that the mass media and other key institutions:

> contribute to the development and maintenance of hegemonic domination [...] They 'connect' the centres of power with the dispersed publics: they mediate the public discourse between elites and the governed. Thus they become, pivotally, the site and terrain on which the making and shaping of consent is exercised, and, to some degree, contested.
>
> (142)

Like numerous news studies demonstrating the distinctive 'authority-orientation' (Ericson et al., 1989) of news media, and the privileged and 'habitual access' (Molotch & Lester, 1974) given to powerful institutions, Hall and his colleagues argued that the practical pressures of news work combined with the key journalistic professional demands of 'impartiality and objectivity' to 'produce a systematically structured over-accessing to the media of those in powerful and privileged institutional positions' (Hall et al., 1978: 58).

While early and now classic studies of news coverage of demonstrations, movements and protest largely confirmed the marginalisation in media coverage of dissenting 'voices' as well as of resource-poor groups and interests (Halloran et al., 1970; Goldenberg, 1975; Gitlin, 1980), the 'hegemonic domination' argument and its particular notion of powerful 'primary definers' has been subjected to considerable criticism and revision. The relatively monolithic notion of hegemonic power articulated by Hall et al. (although itself a vast step forward in relation to earlier, more simplistic Marxian notions of power and dominant ideology) was challenged early on by empirical studies of sources and 'voices' in a range of types of news coverage, including environmental issues (Cottle, 1993), science and medicine/health (Hansen, 1994; Miller, 1999), crime (Schlesinger and Tumber, 1994), the voluntary sector (Deacon & Golding, 1993) and trade unions (Davis, 2002). In a seminal book chapter, Schlesinger (1990) criticised the oversimplification, inherent in the primary definer thesis, of the relationship between media/media professionals and their sources for inadequately accounting for, *inter alia*, contention between official news sources, inequalities of access among privileged sources, and changes over time in source access and status.

Essentially, then, as cogently argued by Schlesinger and shown empirically by numerous studies of source–communicator relationships, media access and definitional power, while certainly highly selective and indeed circumscribed by available economic and cultural resources, cannot be reduced to a simple classification into powerful elites/institutions with privileged access versus resource-poor individuals and groups marginalised by or excluded from the public arena of the news media. In terms of the construction of environmental news and environmental issues as social problems, perhaps the main insight derived from Schlesinger's critique and from studies of environmental news is the highly dynamic and fluid nature of the claims-making process, and of the media's position therein.

We are thus far removed from Hall's notion of powerful institutions and elites acting in relative unison, and instead much closer to notions of a complex – and most of all fluid and constantly changing – system where claims, in Hegelian dialectical fashion, inevitably generate counterclaims, which then synthesise into new claims, provoking new counterclaims, and so on and so forth. In this dialectical process, new fissures constantly open up within the powerful elites and new alliances are continuously formed, just as new and innovative ways of framing claims – making them resonate with cultural climates of opinion – and adjustments to ideological and rhetorical shifts are made. However, it remains important not to equate a more dynamic and complex view of source–media relations or of the claims-making process with a pluralist view that loses sight of the significant structural constraints – including economic and cultural resources – by which they are circumscribed.

EXERCISE 3.1 Contested claims-making/competing claims-makers

There is considerable research evidence to suggest that the news media tend predominantly to turn to and to give privileged access to 'authoritative' and powerful sources in society. Likewise, there is much research evidence (see also further on in this chapter) to show that pressure groups have mixed success when it comes to 'being heard' by the news media. However, even at the relatively simple level of analysing *who* is quoted or referred to and *how often*, the picture may vary considerably depending on, for example, type of issue (e.g. climate change versus genetically modified crops) and the prevailing climate of public and political opinion at the time.

Identify – for example, by searching online news media – a selection of news items about genetically modified crops or genetic modification in agriculture and food production.

Who are the key claims-makers/sources quoted or referred to?

Do environmental pressure groups, individual farmers, 'ordinary' consumers, and so on, 'get heard'? And if so, are their statements or accounts given similar prominence to more authoritative sources, such as politicians, representatives of biotechnology companies, independent scientists, and so on?

Finally, are there any examples of news media highlighting disagreement or clashes within different categories of sources, e.g. different pressure groups disagreeing with each other, scientists disagreeing with each other, major biotechnology businesses disagreeing with each other?

While the claims-makers who often on the surface appear to be most directly and visibly engaged in attempts at managing and influencing media coverage of environmental issues are perhaps environmental pressure groups, there is a growing body of evidence showing that government, industry, business, research institutions and professional associations all increasingly engage in the provision, packaging and management of news. Two areas which have been particularly well documented in the last two decades are those of government news management and political spin generally, and the increasing use of PR by a wide range of not only traditionally powerful and resource-rich elites but also 'outsider' and resource-poor campaigning groups (Davis, 2003). Davis (2003) points to the significant growth in the use of professional public relations practices across the board in what he refers to as 'the new public relations democracy' (p.40).

In his survey of PR and its implications for source–media relations, Davis (2003) argues, resonant with the 'primary definer' criticism delineated above:

> that there is not a simple balance sheet that links media access (or appearances) with favourable media coverage. One cannot simply tally up government, corporate and other source levels of access in order to ascertain who is being more favoured by the media. For example, non-appearance and little obvious media access may be the intention of the source. Frequent appearances in the media are equally likely to be instigated by rival sources and journalists, and can often result in poor media relations and unfavourable coverage. Media access by certain sources is often utilised for the benefit of others. Research on public relations therefore reveals that media-source relations are rather more complex than previous work on sources and news production has assumed. Just as those relations are more complex, so too are the benefits brought by PR to sources.
>
> (40)

In the following, we shall examine in more detail how various sources/claims-makers seek to influence and use the media in their 'construction' of environmental issues as issues for political and social concern. But first it is necessary to emphasise that while the term 'pressure group' appropriately alludes to a deliberate and active approach to influencing and managing public and media discourse, it would be mistaken to assume that all other sources/claims-makers are either simply passive targets for enquiring journalists or indeed necessarily any less 'strategic' in their approach to news media and public discourse.

Claims-making and media visibility

It is tempting to assume that attaining public visibility through media coverage and publicity is the single most important objective for claims-makers in general, and for pressure groups or issue advocates in particular. After all, much media and public opinion research has echoed the notion that 'if you don't exist in the media, you don't exist' (a reference attributed to American journalist Daniel Schorr in Wallack et al., 1999: ix). And it does not merely apply to news coverage and news media. Leading communications researchers have referred to the absence from media coverage as a form of 'symbolic annihilation', a notion which has also been widely used in the prominent strand of communication research known as cultivation analysis (e.g. Gerbner et al., 1994). Cultivation analysis is based around the notion that the media are the important 'storytellers', fictional and factual, of our time, continuously supplying us with a wealth of morality tales about who exists, who is important, who wields power over whom, what is right and wrong, what is acceptable, and so on.

While these arguments are indeed persuasive and fit very well with a constructionist perspective's emphasis on the crucial importance of claims-making – including, not insignificantly, claims-making in public arenas such as the media – it is, however, also necessary to remind ourselves (as discussed in Chapter 2 with reference to Edelman, 1988) that the strategic *management* of publicity – including outright suppression of certain arguments, claims, issues – may be more effective than an all-out endeavour to achieve media and wider publicity for publicity's sake. Wallack et al. (1999) tellingly refer to this in a book, *News for a*

Change, otherwise devoted to strategies for influencing and managing media publicity:

> A while ago, I was completing a presentation on media advocacy when someone from the audience asked, 'But if we get a lot of attention in the media, won't it just mobilize the opposition?' I was very surprised. The question seemed to suggest that remaining invisible was a desirable strategy. I realized that it wasn't the first time I had heard this question, although this was the clearest that it had ever been stated. It seems that for some advocates, being invisible and not drawing attention to their issue is seen as a kind of strategy; they avoid controversy and make only tentative requests for change.
>
> Many others, however, know that power and visibility are important to amplify concerns and advance effective approaches. They know that reticence and invisibility are the problem, not the solution.
>
> (ix)

Widespread publicity may, in other words, have a negative influence on a group's or claims-maker's objectives, particularly if the main effect is to galvanise opposition. As Ibarra and Kitsuse (1993; discussed in the previous chapter) indicate, every claim generates a counter-rhetoric or counterclaim. It is, however, the 'power framework' delineated by Lukes (1974/2005), Edelman (1988) and others which provides the main corrective or qualification to the constructionist perspective's unfettered emphasis on claims-making, namely by stressing that the ability to keep claims and issues *off* the public agenda is just as significant an exercise of power, if not more so, as the ability to successfully place claims and issues on the public agenda or in public view.

A related 'side-effect' of pressure group claims-making is the extent to which (again in the fashion of Ibarra and Kitsuse's argument that every claim generates a counterclaim) it prompts a sharpening of opponents' publicity practices. As Signitzer and Prexl (2007), in their analysis of *greenwashing*, point out, activism pressure often tends to stimulate in corporate organisations what public relations theorists call 'an excellent public relations function' (Grunig et al., 2002), a redoubling of corporate public relations efforts to pre-empt, counter, engage with, accommodate, undermine and frame the arguments and claims of pressure groups. Greenpeace's Brent Spar campaign against Shell (see Box 3.1) illustrates this point, in the sense that it prompted Shell into rethinking their public relations strategy. Similar patterns can be found in relation to most major pressure group-initiated campaigns, but perhaps the most telling example of these kinds of dynamics in our time is in relation to claims-making on climate change, where the strategic communication practices of key stakeholders have changed flexibly and often in response to a clear sense of both the perceived opposition or counterclaims and of shifts in the climate of opinion (Schlichting, 2013; Pollach, 2018).

In terms of understanding how different claims-makers and pressure groups adopt different approaches to media, government and publicity, Grant (2000), in his useful analysis of pressure groups, offers a helpful distinction between *insider* groups and *outsider* groups:

> Insider groups are regarded as legitimate by government and are consulted on a regular basis. Outsider groups either do not wish to become enmeshed in a consultative relationship with officials, or are unable to gain recognition. Another way of looking at them is to see them as protest groups which have objectives that are outside the mainstream of political opinion. They then have to adopt campaigning methods designed to demonstrate that they have a solid basis of popular support, although some of the methods used by the more extreme groups may alienate potential supporters.
>
> (19)

This distinction is useful, not so much for a categorical taxonomy of pressure groups, but for understanding how pressure groups – even within the same general domain or issue area, such as 'the environment' – may adopt widely different approaches to government, the institutions of formal politics, and the media. It is perhaps most useful to see the insider/outsider distinction as not so much a categorical either/or classification, but rather as a continuum, where different pressure groups position themselves at various points, changing over time, between the insider and outsider extremes. Some groups, recognised as legitimate and credible by government, may work most efficiently and effectively behind closed doors, that is, away from the gaze of the media and other public arenas, through the mechanisms of formal political lobbying and institutional processes. Others, like Friends of the Earth, for example, may be formally involved in government consultation processes – and be publicly seen to be involved – on some issues, while, on others, it may be crucial to their publicity strategy to be, and to be seen to be, completely separate from and outside of any process of consultation with government. And others still, most notably Greenpeace, will staunchly maintain an outsider position and strategy, based crucially on a critical (of government and formal institutional politics in particular) and independent stance, unsullied by any hint of collaboration with or co-option by the powers that be.

In terms of media and publicity strategies then, and simply put, insider groups are distinctly publicity-shy, seek to avoid media coverage and publicity, and indeed depend largely for their effectiveness on staying out of the public limelight, on remaining as publicly invisible as possible. Outsider groups, on the other hand, depend – in the absence of formal or direct channels of communication with government and political decision-making processes – entirely on the mass media, on their ability to gain and maintain, in the media and other public arenas, a high profile both for themselves and for the issues on which they campaign. Outsider groups depend on media and related public publicity principally for two reasons: 1) public visibility – and crucially, legitimacy – to help recruit members and financial support for the group's campaigning activities; and 2) as the main channel for achieving public and political attention and action regarding the issues on which pressure groups campaign (e.g. Cracknell, 1993; Cox & Schwarze, 2015).

It is the strategies of these outsider claims-makers or groups – particularly vis-à-vis the media – that we shall examine in more detail in the following section.

Outsider environmental pressure groups and the media

Environmental pressure groups are possibly best known for their ability to make news through spectacular stunts or demonstrations, the most visible aspect of their activities. Indeed, prominent groups such as Greenpeace have traditionally been seen as masters of the art of creating visually appealing news events backed by dramatic film footage made readily available to interested media. But the newsworthiness of environmental pressure groups would soon wear off if they had to rely solely on their creation of spectacular protest 'performances'. Actions such as sailing in small dinghies in front of the harpoons of industrial whaling ships, or sailing underneath the barrels of toxic waste being dropped from large cargo ships, or blocking waste pipes discharging noxious chemicals into the sea are of course eminently newsworthy and visually striking, but they are not sufficient for remaining on the media agenda or for maintaining public visibility in the long term.

One of the main reasons why successful pressure groups have been able to sustain their media coverage is to do with the fact that the major part of their work consists of gathering intelligence about – and drawing media and public attention to – environmental issues that are already being discussed in the forums which the media regularly report on. The most notable and newsworthy of these forums is of course the political forum – the forum of parliamentary and governmental activity.

Theatrical stunts and visually daring protest action of course have inherent newsworthiness, but such actions cannot in themselves explain the long-lasting and sustained accessing of successful pressure groups in media coverage. Nor can such access be explained by additional arguments concerning the skills of such groups in catering to the needs of media organisations in terms of news cycles, provision of sources, provision of visual material, exploitation of deadlines, and so on. These aspects are important, but sustained success as a claims-maker and in terms of getting media coverage requires more. Particularly important is the ability to link to or latch onto developments, events and (important) people in existing established and legitimate news forums.

Most of the issues on which successful pressure groups campaign and successfully gain media coverage are issues which already have an institutional forum rather than completely new issues which have not been problematised in some form or other before. The success as a claims-maker is thus partly explained by careful timing of press releases and publication of reports to coincide with (or often, slightly precede): political events (e.g. debates in Parliament; the publication of government papers/reports); international meetings and conferences (e.g. the World Trade Organisation, WTO, talks; World Monetary Fund meetings; the annual meetings of the most powerful economic powers; and of course most prominently with regard to the environment, the annual Conferences of the Parties, COPs, on climate change); treaty renewals (e.g. the Antarctic Treaty Nations; the treaty on whaling through the International Whaling Commission; the nuclear test-ban treaty); industry or public authority announcements (e.g. announcements of nuclear industry decisions concerning suitable sites for storage of low-level radioactive waste); the publication of independent scientific reports.

In this respect, effectiveness as a claims-maker arises from the exposure of agenda items which are part of the routine and legitimate forums of politics, government, public authorities, international politics, and so on; but while they are routine forums for media attention, it is the claims-making activity of pressure groups that helps direct media and public attention to aspects and interpretations which might otherwise have gone unnoticed or might have been deliberately glossed over. Pressure group claims-making activity then is often at its most effective, not so much in terms of constructing entirely new problems for social and political attention, but in terms of framing and elaborating environmental dimensions which are already in the public domain as issues or problems.

At one level, this is a simple process of gaining coverage by attacking or making claims about people, institutions and forums which are already by themselves newsworthy and the focus of routine interest. At a more complex level, it points to the importance of pressure group 'work' as intelligence gathering and surveillance of developments in environmental policy-making and decision-making. Eyerman and Jamison (1989: 113) aptly refer to this important component of pressure group work as the 'transformation of knowledge into an organisational weapon'. Continuous monitoring (surveillance) of developments in key political decision-making forums and strategic information gathering, followed by strategic framing and dissemination of information, are among the core tasks for successful claims-making, without which spectacular protest actions and demonstrations would soon cease to command attention in the public arena.

Journalists, even specialist journalists (science, environment, health, medical, consumer affairs, and agriculture journalists), could not hope to even begin to monitor systematically the wide range of institutions, industries, bodies, political forums, and so on, involved in decision-making about environmental issues. This requires an organisation or organisations with expertise and resources for precisely such a task. Seen from this perspective, it is perhaps not surprising that successful pressure groups often enjoy a great deal of rapport with journalists. In many respects, they help make the journalists' task easier by offering 'information subsidies' (Gandy, 1982) and perhaps easily digestible interpretations or commentaries on what is often highly complex political processes involving abstruse scientific, economic or other research and data.

Kielbowicz and Scherer (1986: 87) note the importance to media coverage of 'having identifiable leaders authorised to speak for a large following; they seem authoritative, like the head of a large business or government agency'. However, a focus on high-profile, charismatic or celebrity-type leaders or personalities within a pressure group can also backfire. Gitlin (1980), in his analysis of the American student movement of the 1960s, for example, observed how media coverage increasingly centred on personality clashes and internal schisms within the movement, while losing sight of the key issues promoted by the movement. Greenpeace is particularly interesting for its general and comparative success in deflecting media attention away from itself as a pressure group and away from personalities within the group, while succeeding generally in focusing the media and public attention on the issues at hand. This is partly achieved through a highly decentralised structure, where local/regional issues are generally campaigned upon and addressed by local/regional campaigners and spokespersons, which in turn contributes to the organisation's credibility within the local/regional communities immediately affected by the issues campaigned upon (see also Hansen, 1993b). The increasingly global and interactive nature of communication has at once facilitated and amplified the significance and potential of activism that is firmly locally anchored but globally connected (Lester & Cottle, 2015).

An interesting variant type of claims-making is the form where a pressure group is actively *used* by other forums or institutions (or individuals within such institutions) for leaking information and bringing it to the attention of the media. In such instances, the pressure group becomes a conduit or a vehicle for disseminating information, a role which in itself confers legitimacy and prestige on the pressure group.

The more common practice, however, is for a pressure group to attach itself to – and, more importantly, to *frame* – developments in other institutional forums by producing 'evidence' or 'information' which carefully targets a particular aspect of such developments. The evidence generally takes the form of opinion polls, surveys, scientific analyses or studies commissioned by the pressure group, but as Eyerman and Jamison (1989) have pointed out, this is not principally a question of educating the public or producing knowledge or science for the people; rather, it is a question of producing knowledge and information which can be used strategically in public arena debates.

Like advertisers and public relations professionals, successful pressure groups are knowledgeable about the different organisational values, editorial policies, news requirements, political orientation, and – perhaps most importantly – target audiences of different media. Instead of a blunderbuss or shotgun approach to the media, successful groups normally adopt a highly targeted approach, carefully 'packaging' their information and campaigning to suit the particular needs of selected media (Lacey & Longman, 1993). In this respect, they may also seek to cultivate good relationships with selected journalists, who in return for good and fair reporting may be given privileged access to and insights into upcoming campaigns. The careful differentiation of course also caters to the very different visual and textual requirements of different media.

Environmental pressure groups are generally more successful in drawing media attention to particular environmental issues than in gaining coverage for their own definitions of such issues. The mass media are notoriously authority-oriented. Thus, studies of environmental media coverage have virtually without exception shown that the sources who get to be quoted in news and public communication and who get to define environmental issues are – as in most other types of news – predominantly those of public authorities, government representatives, industry and business, and independent scientists and experts. Environmental pressure groups are far less prominent as quoted sources in media coverage (Anderson, 2015; Williams, 2015).

Studies of actors and primary definers appearing in actual media coverage indicate that while environmental groups may be important as initiators or catalysts of public debate or controversy, and subsequent media coverage, they do not, on the whole, command a prominent role in terms of the definitions which are elaborated and contested in the media arena.

While an environmental pressure group such as Greenpeace has generally demonstrated an impressive capacity for securing media coverage for its claims, its capacity to control the way its claims have been framed and inflected by individual media is more doubtful. Individual media thus exercise a considerable amount of 'ideological work', not merely in terms of the differential accessing of sources and selective prominence given to particular sources but, perhaps more significantly, through their differential choice and promotion of particular lexical terms (e.g. Greenpeace as 'terrorists', 'a nuisance', 'undemocratic'), particular discourses (e.g. law and order, democracy, science), and, consequently, particular frames.

The key challenge then for environmental pressure groups and other claims-makers is not so much that of 'commanding attention' (Solesbury, 1976) in the public sphere, but rather that of claiming legitimacy for the problem diagnosis, definition and course of action, that the group or claims-maker is advocating.

Box 3.2

Claims-maker tasks and news management strategies

Three key tasks are set for claims-makers in the construction of social problems (Solesbury, 1976):

1. Commanding attention
2. Claiming legitimacy
3. Invoking action

News management strategies of pressure groups and other claims-makers are:

- Newsworthiness of publicity stunts and demonstrations; but much more is needed for sustaining coverage in the longer term: e.g. intelligence gathering and surveillance.
- Exploiting knowledge of media routines, news values, news cycles, journalistic and editorial preferences, economic pressures on media.
- Selective targeting of media outlets, exploiting news competition between media, knowledge of audience profiles for different media, and enhancing control over framing of coverage.
- Organisational and spokesperson arrangements designed to deflect media interest away from personalities and personality clashes within pressure groups, and to keep media attention focused on the campaign or issue at hand.
- Most pressure/campaign issues are already part of newsworthy institutional forums.
- Careful timing of press releases and commissioned reports in relation to:
    - Planned or scheduled political events
    - International meetings and negotiations
    - Treaty renewals (and similar 'diary items')
    - Industry, business and public authority announcements.

- Framing of items which might otherwise have gone unnoticed:
  - Attacking newsworthy people, organisations or institutions
  - Surveillance of policy-making.
- Helping or 'information subsidising' the under-resourced journalist and media:
  - Selective spotlighting of policy or negotiations
  - Provision of 'ready-made', visually attractive, news footage and commentary
  - Framing through commissioned research which targets developments in other institutional forums
  - Alliance with research/science/academia.
- Issue campaigners and pressure groups are often more successful in drawing attention to broad issues, than in promoting their own specific definitions.

Globalisation, activism and old/new media

Environmental activists and claims-makers have a long and distinguished history of making innovative use of public arenas and associated media technologies. Necessity – limited resources, limited access to mainstream news media, limited access to political decision-makers, and so on – has to a large extent been the mother of invention in this respect. Pressure groups like Greenpeace and Friends of the Earth understood from a very early stage the essential role of visuals (including in the form of the age-old tradition of 'bearing witness') and spectacle (public protest taking the form of symbolic performances) as keys to gaining mainstream media visibility and coverage (see also Rose, 2011, and Doyle, 2007, on the centrality of visualisation in pressure group campaigning). Like their more powerful counterparts in government and public organisations/institutions, successful environmental pressure groups have been skilled in providing what Gandy (1982) refers to as 'information subsidies' to media news professionals and organisations short of both time and resources.

Historically, as new media technologies have become available to wider publics, environmental pressure groups have been quick to adopt and adapt them for their campaigning strategies, including for both external and internal (within pressure group organisations) communication purposes. Thus, fax machines, video recorders and video cameras were quickly pressed into service when they first became widely available in the 1980s. Video-News-Releases, VNRs, became an important way to provide resource-strapped media organisations, hungry for visual material to fill the rapidly expanding number of news programmes and news channels in the late 1980s, with news material. Increasing sophistication of information subsidy practices of pressure groups did not of course lead to unfettered access to mainstream news media where professional journalistic values of objectivity, balance and fact checking dictated the handling of source-provided material (Friedman, 2015; Anderson; 2015).

Email and portable satellite communication equipment capable of virtually instantaneous transmission of evolving 'news events' (including pressure group protest 'performances') came next, and, from the mid-1990s to the present, the exponential growth of the Internet and digital media have likewise been highly enabling technologies quickly adopted by and incorporated into the communication strategies of environmental pressure groups.

One of the early attractions of the Internet and new digital communications for environmental pressure groups and claims-makers was the prospect of altogether bypassing traditional news organisations and news media, instantly overcoming all the associated difficulties of gaining access and of controlling/managing the framing of campaign messages. Combined with developments in visual digital communications technology, the World Wide Web facilitated an unbroken chain of control – by activists themselves – over the orchestration of protest and public performance, the recording and 'packaging' or framing of protest, and the global communication of such news/campaign messages.

EXERCISE 3.2 Measuring pressure group influence on mainstream news media

Identify a recent pressure group campaign, demonstration or similar protest event, that received some coverage in a selection of mainstream news media (e.g. in major online newspapers, or major online news of broadcast organisations such as the BBC (http://news.bbc.co.uk)). Compare this coverage with the pressure group's own publicity and accounts of the action on its own website or through material placed by the pressure group on YouTube (http://www.youtube.com) or other websites.

Is the overall 'message' of the mainstream media coverage similar to or significantly different from the 'message' conveyed by the pressure group's own communications?

In order to address this question, focus, for example, on some or all of the following questions:

Are there recognisable visual or verbal sequences in the mainstream news coverage that clearly originate from the pressure group's own communications?

What particular aspects do the mainstream news media select and focus in on in their coverage? Are these emphases different from those of the pressure group's own communications?

Who (e.g. scientists) or what (e.g. published research) is quoted or referred to as 'authoritative' sources in the mainstream news and in the pressure group's communications?

What is the balance in media coverage and in pressure group communications between a focus on events (e.g. clashes with the police or other authorities, damage to property) and a focus on issues (e.g. pollution, energy policy, climate change)?

Are there marked similarities or differences in the types of words (e.g. adjectives, metaphors, etc.) that are used?

Are there marked similarities or differences in the visuals that are used (e.g. photographic angles and frames)?

Your answers will give some idea of the extent to which pressure groups can manipulate or influence mainstream news media coverage, and they will give an indication of whether the pressure group can be said to have successfully communicated its message to/through the mainstream news media.

An early and much celebrated example of the new opportunities afforded by the digital communications environment was that of the anti-globalisation protest around the World Trade Organisation summit in Seattle in November of 1999. This provided one of the first potent examples of how the Internet and new communications technologies were used for rapid information exchange and news framing detached from the central control of mainstream media and political authority, and for organising and orchestrating disparate protest groups into coordinated action (DeLuca and Peeples, 2002; Wall, 2002; Cottle, 2006).

Away from the glow of rousing enthusiasm about the supposed vast (democratic) possibilities of the Internet and related new communication technologies – much of which is little more than a repetition of the enthusiastic fanfare of hopes for a democratisation of communication with which every new communications technology has historically always been greeted – a growing body of research on the use of new media by social movement organisations, pressure groups and indeed claims-makers in general (see e.g. Hestres and Hopke, 2017) has shown a rather more nuanced and indeed less sweepingly enthusiastic picture – at least with regard to democratisation of the public sphere.

Initial euphoria about the implications of the Internet's collapsing of space and time barriers (i.e. virtually unhindered instantaneous communications access across geographical, national, cultural and political boundaries and divides) for movement and pressure group mobilisation have been tempered by the persistence of a 'digital divide' along traditional demographic and geopolitical lines. By 2015, still less than half of the global population had access to the Internet from home, and considerable disparities continued across regions and countries of the world, as well as within countries along traditional socio-demographic lines (Segerberg, 2017). However, this is clearly a fast-changing field where not only does access continue to increase, but continuous technological development opens up new communications opportunities and possibilities.

The change underway since the late 20th century from a public communications sphere dominated by print and broadcast mass media to the new digital communications environment has offered exciting new opportunities that have on one level complemented, augmented and significantly enhanced traditional environmental campaigning, claims-making and communication, while at the same time enabling new forms of movement and protest organisation.

Traditional environmental pressure groups and NGOs were quick to take advantage of the key features – time/space compression and comparatively low cost – of new digital communications technologies to augment and enhance: 1) *internal organisation, management and communication*; 2) *protest and action planning and coordination*; 3) *communication with members as well as recruitment of new members*; 4) *campaigning and discursive/rhetorical 'sparring' in the public sphere of the Internet*; 5) *campaigning and identity building, including the framing and boundary drawing that is associated with website design, messaging and linking practices.*

But perhaps more significantly, the new digital communications technologies have enabled and given rise to new forms of claims-making, protest, organisation and mobilisation. As summarised by Hestres and Hopke (2017):

> In some cases, digital communication technologies have simply made the collective action process faster and more cost-effective for organizations; in other cases, these same technologies now allow individuals to eschew traditional advocacy groups and instead rely on digital platforms to self-organize. New political organizations have also emerged whose scope and influence would not be possible without digital technologies. Journalism has also felt the impact of technological diffusion. Within networked environments, digital news platforms are reconfiguring traditional news production, giving rise to new paradigms of journalism. At the same time, climate change and related issues are increasingly becoming the backdrop to news stories on topics as varied as politics and international relations, science and the environment, economics and inequality, and popular culture.
>
> (1)

From a media and communications perspective, perhaps the single major implication of new digital and mobile communications technologies for the public construction of the environment as a social problem is the twin emergence and mass proliferation of voices/claims-makers about the environment combined with the concomitant erosion of control (including the control exercised by traditional news media) over news and information about the environment, environmental problems, environmental damage – with the associated public assignment of responsibility – and environmental protest.

The new digital media and communications landscape, then, offers greatly enhanced opportunities to contribute to the public communication about the environment, and in this sense it also provides multiple public arenas for competition between key interests in society. These developments in the media and communications landscape thus, as Boykoff et al. (2015: 227) note, 'prompt us to reassess boundaries between who constitute "authorized" speakers (and who do not) in mass media as well as who are legitimate "claims-makers"'.

As shown earlier in this chapter, one of the hallmarks of successful environmental pressure groups is their ability to combine newsworthy, eye-catching publicity stunts with careful intelligence gathering and packaging, and the ability to serve up carefully researched intelligence/ information in a form that exploits the news values, journalistic routines and organisational arrangements of mainstream news organisations. In a news-gathering environment characterised by ever-tightening economic pressures, the increasing ability to gather news via the Internet without ever leaving the newsroom must clearly have major implications for how environmental correspondents and other media professionals covering the environment go about their work. We shall examine this further in the next chapter.

Another important dimension, but one that has received comparatively less attention so far, concerns the credibility implications of news and news-source proliferation on the Internet. If mainstream media/news organisations, who have traditionally been seen as credible and reliable conveyors of news and have commanded a relatively high degree of trust from their audiences, are increasingly bypassed by or in direct competition with a wealth of other sources offering different accounts of events, arguments and debates, then how does this affect the way in which news is consumed and assessed by various audiences/publics? If in cyberspace every account or opinion is traded equally, then how do audiences go about determining what is credible and reliable and what isn't? The ecology of how we as publics and citizens draw on and interact with mediated environmental communication is changing rapidly: although the major news media continue to play a major role, environmental news is now generated by a much wider range of providers, including organisations dedicated to the online provision of environmental news, information and comment (Friedman, 2015), freelance environmental journalists (Brainard, 2015), and the rise of citizen journalism (Allan & Ewart, 2015) – we shall explore these developments further in the next chapter. Concomitantly, the way in which we, as publics and citizens, consume, engage with, react to and are influenced by news and information about the environment is changing (Roser-Renouf et al., 2015; Lewandowsky et al., 2017) – we shall explore these changes further in Chapters 7 and 8.

Corporate image strategies

As indicated several times in the above discussion of environmental pressure groups and their media management and publicity strategies, such groups are of course not alone in using communication strategies as a means of influencing, framing and manipulating public awareness, opinion, discourse and action with regard to environmental matters. Far from it. Nor indeed are they necessarily the most effective, and rarely can they call on the kind of economic and political resources available to their key opponents or competitors in the sphere of public claims-making, notably big business and industrial corporations and governments. Business, industry, government and their associated institutions and representatives make use of many of the same PR and media management strategies – increasingly referred to as 'strategic communication' (Frandsen & Johansen, 2017) – deployed by environmental pressure groups. In doing so, they are advantaged by the ability to bring far greater economic and political resources to bear, in terms of campaigning and lobbying, than environmental 'outsider' groups.

While much of the most effective 'campaigning' and 'advocacy' work of governments and industry/business takes place through formal political channels, institutions and lobbying processes, strategic communication and public 'image management' more generally have increasingly become central and crucial dimensions, not least as a response/reaction to the claims-making activities of environmental pressure groups. Many of the major environmental, scientific and health-related controversies of the past decades can thus be characterised to a large extent as discursive or rhetorical public contests/battles fought in terms not of right or wrong, or on the grounds of scientific evidence, although science plays an important and prominent role in the construction of claims, but in terms of 'image', culturally resonant arguments, fear tactics, 'spin', style and 'slick' campaigning.

In an early example, the oil company Shell was taken by surprise at the impact of Greenpeace's Brent Spar campaign in 1995, surprise perhaps mainly that Greenpeace's campaign succeeded in stirring up public and political unease, and, in some countries, public protest actions against a company that had very carefully 'played by the book'. Shell had thus carefully gone through all the necessary scientific research and arguments, and through all the legally required processes to arrive at the course of action subsequently resisted by Greenpeace and seemingly by large sections of the public. Shell, by their own admission, recognised that they had perhaps been too focused on the science and the formal legal process, and had failed to adequately consider 'the hearts and minds' of the public, as the then head of Shell UK, Dr Chris Fay, expressed it in a BBC2 documentary broadcast as the controversy unfolded. Shell very quickly remedied its brief and intermittent lapse of concentration on the public image battle through a combination of traditional image management strategies, including PR, clever website usage and image advertising.

While press releases continue to be an important part of the media and public sphere strategies of business and industry, corporations are also acutely aware that mainstream news organisations – or more specifically, the journalists and media professionals working in these organisations – are weary of what they tend to regard as a tiresome flood of thinly veiled promotional propaganda. Most of all, deep-seated professional criteria of objectivity and impartiality dictate that major media organisations and professional journalists have to be seen to be above giving a platform to the promotion of particular interests. But then, press releases are just one amongst an increasingly diverse range of communications forms available to corporations, governments, NGOs and others involved in strategic communication in the public sphere. Given their often abundant economic resources, large corporations often use image or issue advertising as a key strategy for influencing public /political opinion. Pressure groups of course do so too, but the key difference is one of available economic resources for this highly expensive strategy.

Corporate environmental communication can be characterised under the three categories of product advertising, image enhancement and corporate image repair (Cox & Pezzullo, 2016). The strategic use of discourses and images which associate 'green', 'pure', 'nature/natural', 'sustainable', 'environmentally friendly' qualities with commercial products has, not surprisingly, become an increasingly central and effective feature of commercial advertising (see also the discussion in Chapter 6). In parallel, corporations, businesses and public institutions have increasingly used strategic communication for the purpose of enhancing their environmental image and credentials. This may entail publicising genuine corporate efforts towards more sustainable environmental practices, or it may entail reframing and repositioning public discourse and understanding with regard to a particular industry's environmentally 'tarnished' image, as examined for example in environmental communication research on the coal industry and its communication practices to reposition coal energy as a 'clean' form of energy (Schneider et al., 2016).

In her incisive critique of 'the corporate assault on environmentalism', Sharon Beder (2002) points to the important strategy often adopted by major corporations, namely that of creating lobbying or publicity groups or organisations that can do the necessary claims-making, promoting the interests of the corporations behind them, while at the same time appearing to be relatively independent and impartial participants in the public sphere debate. An absolute cornerstone of the public sphere ideal (Habermas, 1989) and of modern parliamentary democracy is of course the clear and transparent separation of individual (or corporate) commercial interests from public and political processes aimed at ensuring the advancement of the 'common good', of what is in the interest of society as a whole rather than of the economically powerful. The power of this ideal is reflected in (healthy) public scepticism of anything that smacks of propaganda or the thinly veiled promotion of particular corporate or commercial interests. Perhaps more significantly for the present discussion, it is reflected in deep-seated journalistic professional antipathy and resistance to anything that looks like attempts at manipulating news coverage and publicity for commercial or corporate gain. It is in this context then that we can understand why it is important – and potentially far more effective than direct campaigning – for business and industry to do their claims-making through seemingly independent 'front' groups.

A prime and much cited example in the environmental field is the Global Climate Coalition (GCC), 'a coalition of fifty US trade associations and private companies representing oil, gas, coal, automobile and chemical interests' (Beder, 2002: 29), whose prime objective was to cast doubt on the evidence for global warming/climate change and to fight scientific, political and legislative initiatives to curb carbon dioxide emissions. The key strategy (frequently referred to in the critical literature as climate change scepticism, denial or contrarianism, and well tested and effectively used in the tobacco industry's long fight against legislation to curb the promotion and use of tobacco) was to cast doubt on the scientific evidence and consensus on climate change (Oreskes & Conway, 2010) and to oppose attempts to regulate or restrict the fossil fuel-dependent industries.

While, as Miller and Dinan (2015: 90) point out, the GCC succeeded in opposing and delaying legislation to address climate issues and climate change in the early 1990s, they also importantly note how a number of the original members of the GCC made strategic choices, including in terms of their communications practices, to pursue a more subtle approach than that of casting doubt on the increasingly robust evidence emerging from climate science. This approach revolves around accepting and engaging with the scientific consensus and taking leadership in influencing public debate and policy in a direction compatible with business interests. Miller and Dinan (2015: 90) refer to this approach as the 'attempted corporate capture of environmental policy' and at its core is the strategy of corporations repositioning themselves, not least in terms of their public image, as responsible and enlightened corporate citizens, leading the way (through science and innovation) in terms of addressing/finding solutions to climate change.

A specific instance of corporate image strategy is that of 'image repair' following a corporation's involvement – in terms of direct or indirect responsibility – in a major environmental scandal or disaster, such as the VW emissions scandal of 2015 or the BP Deepwater Horizon oil spill of 2010. In an innovative study of BP's strategic framing of communications following the Deepwater Horizon oil spill, Schultz and her colleagues (2012) – analysing BP press releases, and news articles in US and UK newspapers – show that BP successfully deployed a 'decoupling strategy' to dissociate itself from responsibility for the causes of the oil spill crisis, while casting itself as the provider of solutions both to the immediate crisis and to the longer-lasting restoration necessary.

In a comprehensive meta-analysis of studies of corporate and industry communications on climate change between 1990 and 2010, Schlichting (2013) demonstrates the astute changes and adaptations – for example, adapting to changing climates of opinion and political contexts – of strategic corporate communication. Her analysis documents the strategic communication practices pursued by large corporations across different countries in their effort to influence public discourse and political/legislative responses regarding climate change in ways advantageous to their business interests. Significantly, the analysis shows how stance and approaches of strategic communication are flexibly adapted to a changing context. Thus, over the 20-year period studied, industry actors used three dominant frames to define the meaning of climate change: an *uncertainty* frame aimed at casting doubt on and disputing the emerging scientific evidence about anthropogenic climate change (early and mid-1990s; mainly in the USA); a *socio-economic consequences* frame invoking the potential cost to social wealth and living standards of increased legislation to curb climate change; and an *industrial leadership* frame, casting industry actors as taking responsibility and taking the lead in providing (technical) solutions to combat climate change. The analysis demonstrates key variations across different regions/countries and the significant change over time from the dominance of an obstructive *uncertainty* frame to the dominance of a progressive *industrial leadership* frame, but it also importantly notes that all three frames continue to be present throughout. As in the case of the climate change frames identified by Schlichting, it is also safe to assume that the various different strategic communication approaches (direct advertising; the use of front groups, think tanks and coalitions of various sorts; social media use; and message strategies focused on sowing doubt, repositioning of the debate or taking leadership) will continue to feature prominently in public sphere contests over the environment.

Conclusion

News about the environment, environmental disasters and environmental issues or problems does not happen by itself but is rather 'produced', 'manufactured' or 'constructed'. Environmental news, like other types of news, is the result of a complex set of interactions between claims-makers and media organisations and their operatives. The way these operate is in turn circumscribed by political, economic and cultural factors. Building on the argument of the previous chapter, that environmental issues or problems only become recognised as such through the process of claims-making, this chapter has examined the communication strategies and influence of key claims-makers – including environmental pressure groups, government and business/industry – in environmental debate and controversy.

Key tasks for claims-makers were identified as those of commanding attention, claiming legitimacy and invoking action. These tasks in turn serve as equally useful focal points for the assessment of the effectiveness or success of any claims-making activity or campaign. While extensive news media coverage may often be seen as the single most important objective for pressure groups and other claims-makers aiming to get their definitions into the arena of public debate and aiming to influence public and political opinion, it was also noted that an equally important exercise of power is the power to keep issues off the public agenda, away from public scrutiny, with a view to avoiding the mobilisation of opposition or 'counter-rhetorics'.

Key strategies used by pressure groups and other claims-makers for managing and influencing news organisations were examined, including timing, differentiated targeting of media, intelligence gathering for the purpose of 'information subsidising' news media, latching onto or piggybacking on issues or people already visible in important news forums, and the 'alliance with science' – that is, evidence-based argumentation.

New information and communication technologies have been adopted by claims-makers from all sides, and the Internet in particular offers new ways of movement organisation and campaigning, new modes of image management, new ways of engagement with competitor claims-makers, new ways of discursive demarcation and new scope for transgressing or collapsing traditional geographical and temporal communication obstacles and boundaries. Rather than replacing or completely revolutionising traditional modes of communication, new communication technologies have been incorporated as additional layers and complementary to traditional media and communications forms. But they have also facilitated the emergence and effectiveness of entirely new political organisations and advocacy groupings. And they have had, as we shall see in the next chapter, a profound influence on news organisations and environmental journalism.

While much of the communications research literature has focused on the media and news management strategies of pressure groups, it is important not to lose sight of the equally active – and economically far better resourced – public relations and image management strategies of government and corporate business/industry. The final section of the chapter thus briefly examined corporate strategic communication, noting the use of digital/social media, front groups, think tanks and coalitions of various sorts, and strategies of 'sowing doubt' and muddying the waters of public understanding of environmental issues such as climate change, strategies of 'image repair/enhancement' deployed to reposition corporations/industry/business as responsible citizens and innovative leaders in finding solutions to fixing environmental problems.

One of the characteristics of what German sociologist Ulrich Beck has called the 'risk society' (Beck, 1992) – and of postmodern society – is the increasing erosion of public trust in scientific and political authority. This has direct implications for claims-making practices, which are now much less about invoking single authoritative epistemologies (e.g. science, religion) and to a greater extent about the management and manipulation of images, emotions and arguments.

Further reading

Cox, R., & Schwarze, S. (2015). The media/communication strategies of environmental pressure groups and NGOs. In A. Hansen & R. Cox (Eds.),

The Routledge Handbook of Environment and Communication (pp. 73–85). London and New York: Routledge.

Cox and Schwarze provide a state-of-the art overview of research on environmental groups' uses of media, including digital media, and other modes of communication that are intended to generate publicity. They survey such groups' communicative practices, cultivation of news media, rhetorical appeals, issue frames, and alignment of media and audiences, and adaptation to the digital media landscape.

Frandsen, F., & Johansen, W. (2017). Strategic communication. In C. R. Scott & L. K. Lewis (Eds.), *The International Encyclopedia of Organizational Communication* (pp. 2250–2258). Oxford, UK: Wiley-Blackwell.

Mapping the diverse and varied definitions of strategic communication, Frandsen and Johansen show the concept's increasing popularity, from the mid-2000s onwards, among both practitioners and academics.

Hestres, L. E., & Hopke, J. E. (2017). Internet-enabled activism and climate change. *Oxford Research Encyclopedia of Climate Science*. Retrieved 2 Oct. 2017, from http://climatescience.oxfordre.com/view/10.1093/acrefore/9780190228620.001.0001/acrefore-9780190228620-e-404

The authors examine how digital communication technologies have made the collective action process faster and more cost-effective for organisations, while also allowing individuals to eschew traditional advocacy groups and instead rely on digital platforms to self-organise. They show how digital communication technologies are changing the types of advocacy efforts that reach decision-makers and may influence policies. Widespread adoption of digital media, they argue, has fuelled broad changes in both collective action and climate change advocacy. They offer examples of advocacy organisations and campaigns that embody this trend such as 350.org, the Climate Reality Project, and the Guardian's 'Keep It in the Ground' campaign.

Miller, D., & Dinan, W. (2015). Resisting meaningful action on climate change: think tanks, 'merchants of doubt' and the 'corporate capture' of sustainable development. In A. Hansen & R. Cox (Eds.), *The Routledge Handbook of Environment and Communication* (pp. 86–99). London and New York: Routledge.

Miller and Dinan examine the ways in which corporate and policy elites organise and communicate strategically in relation to climate change. They explore specifically the role of think tanks, institutes and other lobbying organisations that have played a significant role in public relations and campaigning on climate change.

Rose, C. (2011). *How to Win Campaigns: Communications for Change* (2nd ed.). London: Routledge.

Chris Rose, a veteran environmental campaigner who has worked for Greenpeace, Friends of the Earth, WWF International and other organisations, offers a wealth of practical insights into what works in campaigning, and strategies for the planning, execution and evaluation of campaigns.

Segerberg, A. (2017). Online and social media campaigns for climate change engagement. *Oxford Research Encyclopedia of Climate Science*. Retrieved 23 Aug. 2017, from http://climatescience.oxfordre.com/view/10.1093/acrefore/9780190228620.001.0001/acrefore-9780190228620-e-398

Segerberg charts the emergence of online campaigning and explores the role of online and social media in how campaigners render the issues and pursue their campaigns, how publics respond, and what this means for the development of the broader public discourse. She identifies core debates concerning the capacity and impact of online campaigning in the areas of informing, activating and including publics.

The environment as news

News values, news media and journalistic practices

THIS CHAPTER:

- Focuses on the roles and 'work' of the media and media professionals in communicating environmental issues.
- Discusses how research on news values, on organisational structures and arrangements in media organisations, and on the professional values and working practices of journalists and other media professionals can help explain why some environmental issues become news, while others do not; why some environmental issues become issues for media and public/political concern, while others fail to do so.
- Examines the development and significance of specialist environmental journalists and of the 'environment beat' in the creation of environmental news coverage.
- Discusses whether journalism is becoming increasingly reactive, rather than proactive, and the impact of new information and communication technologies – and associated new source publicity and communication practices – on journalistic work in reporting on environmental issues and controversies.
- Examines the impact of key journalistic values, such as objectivity and balance, and the strategies environmental journalists deploy in order to cope with the scientific uncertainty which often characterises environmental issues and problems.
- The chapter ends with a discussion of the limitations of the sociology-of-news framework, and the ways in which some of these limitations have been addressed through perspectives focusing on cultural resonances in the discursive construction of environmental issues.

Introduction

News coverage of the environment and environmental issues is the result of complex processes of 'construction' (the literature on how news comes about also deploys a number of other and similar terms – 'production', 'manufacture', 'packaging', etc. – to indicate that news is the result of active 'work'), rather than something that just happens by itself or as a result of obvious observable events, accidents or disasters. If we accept the notion of environmental news coverage as actively 'constructed' – and concomitantly and emphatically reject the models of news media commonly referenced in such metaphors as 'a mirror of reality/society' or news as 'a window on the world' – then what becomes interesting from a communications and sociological point of view is to examine the key operatives/agents, the key institutional settings/forums, and other factors which circumscribe and impinge on the processes of news construction. And we need, of course, to appreciate how the traditional dynamics of the 'production of news' are continuously changing in a fast-evolving media landscape.

In the previous chapter, we started this process by looking closely at the publicity and communications practices of some of the principal claims-makers in environmental debate and controversy. In this chapter, we extend the focus to the media and media professionals themselves. The chapter discusses how research on news values, on organisational structures and arrangements in media organisations, on the professional values and working practices of journalists and other media professionals can help explain why some environmental issues become news, while others do not; why some environmental issues become issues for media and public/political concern, while others fall by the wayside.

The environmental news beat and environment correspondents

When scientists, pressure groups and others first started drawing attention to environmental issues in the 1960s, one of the problems they faced in terms of getting media coverage for these concerns was the simple problem that newspapers and other media did not have an obvious category or rubric for these; they did not have an environmental beat or specialist reporters whose task it was to report on the environment. Thus, it was often not clear whether an environmental issue was a job for the medical or health reporter, or for the science correspondent, or for the political or economic correspondent, and so on (Schoenfeld, et al., 1979).

This is, of course, not to say that the environment did not receive media coverage prior to the 1960s; indeed, several historical studies

(Krieghbaum, 1967; LaFollette, 1990; Nelkin, 1995) have documented the role of science and technology correspondents in particular in covering issues and events that might now be considered as 'environment' stories. But, as Schoenfeld et al. (1979) argue, the 1960s saw the rise of a new and more holistic approach to the environment in the form of 'ecology', one that argued for the need to see everything as connected to everything else, a holistic perspective that very much departed from a view of environmental disasters, events or incidents as isolated occurrences. However, for this kind of perspective to 'fit' with the routine divisions and organisation of news work, it was necessary to create a new specialist category for 'environmental reporting'.

With the creation of specialist environmental beats and the appointment of environmental correspondents towards the end of the 1960s, the media at least became geared up for covering these complex issues which often straddle several subject domains such as science, health, politics and the economy. The appointment of environmental correspondents meant that individual media had reporters who could be on the look-out for environmental news, and this in turn facilitated a marked increase in the amount of coverage.

The importance of specialist environment correspondents to the amount of coverage given to environmental matters in the media seems, perhaps not surprisingly, to be directly reflected in the cyclical trends that characterise long-term media attention to environmental issues. Friedman (2004), in her overview of American environmental journalism, notes that:

> the environmental beat has never really been stable, riding a cycle of ups and downs like an elevator. These cycles, and consequent increases or decreases in numbers of environmental reporters and their space or air time, appear to be driven by public interest and events, as well as economic conditions.
>
> (177)

The rapid growth in environmental coverage in the British media towards the end of the 1980s – a growth that was to a large extent driven by developments in the political arena, and very particularly by the then prime minister Margaret Thatcher's appropriation of 'environmental issues' as central to the Conservative government's policies – brought with it a rapid expansion of environmental correspondents, including in the British tabloid papers, few of which had a dedicated environment beat prior to 1988.

Many newspapers and broadcast media which were quick to set up environmental news beats and to appoint specialist environment correspondents in the late 1980s were however equally quick to drop these posts as the environment slipped down the political priority list in the early 1990s (Gaber, 2000).

Friedman (2004) charts the further development of American environmental journalism in the 1990s, describing the 1990s as the decade that environmental journalism:

> grew into its shoes, becoming more sophisticated with the help of the Internet and a professional organisation, the Society of Environmental Journalists. The field also matured as stories changed from relatively simple event-driven pollution stories to those of far greater scope and complexity such as land use management, global warming, resource conservation, and biotechnology. Growing into shoes can be painful if they pinch, however, and environmental coverage, like most other journalism faced a shrinking news hole brought about by centralisation of media ownership, revenue losses and challenges from new media. Environmental journalism's dilemma was dealing with a shrinking news hole while facing a growing need to tell longer, complicated and more in-depth stories.
>
> (176)

Some of the points made by Friedman are also echoed in the findings from a comprehensive national study by Sachsman et al. (2006) of US environment reporters. Sachsman and his colleagues found, *inter alia*: that a shrinking news hole was seen as one of the top barriers to environmental reporting, and a greater barrier than interference by editors; that 'newspapers were far more likely than television stations to have a reporter covering the environment on a regular basis' (p.98); that 'the use of environment reporters tended to increase along with the size of the 550 newspapers examined' (p.98); that 'most of the environment reporters (...) were veteran journalists' (p.101); that autonomy in story selection was among the top-rated factors among environment reporters; that they relied more often on local and state sources than on national sources; and that most 'felt the need to remain objective, rejecting calls for advocacy or a civic-journalism approach' (p.93).

The pressures on environmental journalism have continued unabated into the 21st century, and the nature and organisation of environmental news and journalism have been transformed by rapid changes experienced since the 1990s in the media and communications landscape. Surveying these developments, Friedman (2015) notes how media convergence, downsizing and the rise of the Internet and digital media technologies have caused the reduction or elimination across traditional news media of specialist environment beats and designated environmental reporters. Environmental news is thus increasingly dispersed across more mainstream news categories and covered by general rather than specialist environmental reporters, with potential implications in turn for the quality, accuracy or investigative nature of reporting on the environment. Sachsman and Valenti (2015) confirm these trends when they cite figures from The Society of Environmental Journalists (in the USA) showing a near-halving of the number of environmental newspaper reporters in the first decade of the 21st century.

While the pressures on specialist environmental journalism and particularly the decline in environmental news beats have been well documented for North America and Europe, it is also important to note that similar trends do not necessarily apply in the same way or to the same extent in other parts of the world. Bauer et al.'s (2013) comprehensive global survey of science journalists – whose remit often includes the environment and environmental issues – thus notes how science journalism in Latin America, Asia and Africa has not been affected by the crisis concerns felt in North American and European science journalism. Recent work on environmental journalism in China (Tong, 2015) and in South America (Pinto et al., 2017) likewise indicate different trends and different types of pressures in the construction of environmental news.

Research on environmental journalism (and science/health journalism) has often revolved around the notion that environmental journalism/news is 'different' from other types of journalism/news in that environmental journalists and editors are more likely to be positively disposed towards their subject matter, and towards sources or claims-makers critical of the status quo, than, for example, political reporters or crime reporters. In an early study of environmental journalism in the UK, Lowe and Morrison (1984: 82) thus found that 'In interviews, environmental journalists expressed undisguised sympathy for many of the issues raised by environmental groups.'

While journalistic and editorial sympathy towards environmental issues promoted by environmental interest groups has similarly been documented elsewhere (Schoenfeld, 1980; Porritt & Winner, 1988; Linné & Hansen, 1990; Sachsman et al., 2010; Bruggemann & Engesser, 2014), the evidence from analyses of media coverage of environmental issues essentially confirms – at least in traditional news media – the authority orientation characteristic of news reporting in general. In environmental news coverage authority-orientation thus manifests itself both in terms of 'indexing' (Bennett, 1990; Shehata & Hopmann, 2012) to official/government viewpoints and positions, and in terms of reliance on recognised official sources/institutions, rather than on, for example, environmental pressure groups or other non-governmental organisations. The authority orientation – and particularly the indexing to government positions – has also been found, not surprisingly given the more authoritarian political systems, in countries such as China (Xie, 2015) and Russia (Poberezhskaya, 2014).

In terms of 'difference' from other types of journalism, studies of environmental reporters and specialist correspondents in the closely related fields of science, technology, health and medicine have shown, *inter alia*, that environment correspondents tend to remain much longer with their specialism than other types of journalist (Hansen, 1994; Friedman, 2004; Sachsman et al., 2006). They are more likely than other journalists to have a science degree or indeed any university degree (Sachsman et al., 2010; Bruggemann & Engesser, 2014), although this is rarely seen by the journalists themselves as a particular advantage in their day-to-day task of reporting – a recurrent refrain from environmental journalists themselves is that they are *journalists first* and *environment/science/medical correspondents second* (Hansen, 1994; Hargreaves & Ferguson, 2000; Sachsman & Valenti, 2015). They often have more contact with fellow environment/science correspondents in competitor media than they do with colleagues in their own medium or organisation – referred to in the literature as a 'competitor–colleague' relationship or as the 'inner club' (Dunwoody, 1980). They have – and value having – a greater degree of autonomy from editorial interference than general reporters; that is a greater degree of freedom to decide on what to cover and how, although it is also the case that the autonomy of the environmental/science reporter varies considerably depending on the size (e.g. national versus regional or local) of media organisation and on the type of media organisation (i.e. 'quality' or 'elite' media versus 'popular' or 'tabloid' media; Hansen, 1994; Sachsman et al., 2006). Crucially, the degree of autonomy in environmental journalism is increasingly under pressure and being eroded by the overall changes in and pressures on the traditional media – particularly for far fewer environmental reporters to do far more reporting (Brainard, 2015; Williams, 2015).

Finally, there is a relatively prominent argument in the literature on science and environment journalism that specialist correspondents in these areas tend to develop a deferential and uncritical relationship with their sources in what has been called a 'symbiotic' interdependency (Friedman, 1986; Goodell, 1987; Nelkin, 1995). The evidence that a relatively close relationship between specialist environmental correspondents and their sources leads directly to a deferential, celebratory or uncritical type of reporting is, however, not particularly strong or convincing. What we do know from a growing body of research on environmental journalists is that there is indeed considerable evidence to suggest that the development of a comprehensive network of 'reliable' and 'trustworthy' sources is seen as a key component of becoming a good reporter (Anderson, 1997; Hansen, 1994; Sachsman et al., 2010; Bruggemann & Engesser, 2014). It is also an essential component of the specialist correspondent's strategy for dealing with the uncertainty that often characterises environmental issues, as well as with the controversial and contradictory nature of evidence in public debate about environmental problems (Dunwoody, 2015).

Proactive/reactive journalism

Much can be learned about the production of environmental news by looking more closely at the practices of environmental journalists. It is perhaps tempting to think of the traditional image of the journalist as constantly on the look-out for a news scoop, continuously scouring the key news forums for new information, and vigilantly telephoning a large array of news sources for information, leaks and news about developments. Of course, these are all aspects of journalistic work, but as the media and communications environment has expanded and radically changed, journalistic work has also changed towards a desk- or screen-bound culture of *reacting* to the masses of information now permeating the digital communications environment.

Rather than roaming around news conferences, telephoning sources, mingling with key scientists and decision-makers, the role of the journalist today is often one of monitoring the digital communications environment, sifting through and reacting to the flow of information constantly generated by a large array of sources, from individual citizens through to research institutions, corporations, businesses, NGOs and governments. A direct implication of this change in the nature of news work is that those with the largest publicity and communications resources, other things being equal, stand a much better chance of gaining access to the news than those who have few or no resources for mobilising and pushing information and definitions to the news media.

'Publicity resources' in this context does not simply refer to the volume of communication; it refers crucially to the ability of sources to package and frame information in ways which resonate well with the perspectives and ideology of the news media, and, at a more direct level, ways which make the journalist's news-writing task easier:

> news is framed by the sources who have the most access to journalists, and who provide a socially constructed interpretation of a given set of events or circumstances that makes a journalist's job easier by providing an acceptable structure for the ensuing news stories. These sources exercise social and political power by steering journalists toward one particular self-serving way of framing the story.
> (Smith, 1992: 28)

There is nothing new about the concern that news sources may have a great deal of influence on what becomes news. Gans (2004 [1979]), in his classic sociological study of news organisations, elegantly described the relationship between sources and journalists as resembling 'a dance, for sources seek access to journalists, and journalists seek access to sources' (p.116) and he went on to indicate that this was not an equal partnership: 'Although it takes two to tango, either sources or journalists can lead, but more often than not, sources do the leading' (p.116).

In a pioneering study of environmental news reporting and source influence, Sachsman (1973) notes that environment reporting in the 1940s often was dominated by corporate public relations efforts. 'By the late 1960s, however, (...) environment reporting was based on conflicting statements from a wide variety of sources, ranging from environmental activists to government officials and business leaders' (Sachsman et al., 2006: 95). Sachsman's (1976) study of environmental news sources and information used by media in the San Francisco Bay Area found that over half of environmental news reports originated in or drew directly on source-generated press releases and public relations efforts. Sachsman also found that in many cases news reports amounted to little more than a minor rewriting of the press releases that gave rise to the news coverage.

A long tradition of studies from North America, Europe and Australia have similarly confirmed the tendency for environmental news reporting to rely very predominantly on government and 'authoritative' institutions, on scientists and independent experts, rather than on non-governmental groups (NGOs) or indeed on environmental pressure groups. More recently, and confirming the long-accepted notion that basic journalistic practices and values differ little across the globe, similar findings have begun to emerge in studies of environmental journalism in Russia (Poberezhskaya, 2014), China (Tong, 2014; Xie, 2015), India and countries in both Latin America (Pinto et al., 2017) and Africa.

EXERCISE 4.1 Who sets the news agenda on environmental issues?

Take a look at online environmental news items in major news media on any given day. For each news headline, ask: 1) why is this 'news' on this particular day? Whose activity in which news forum has caused this to be considered newsworthy by this news medium on this particular day? And 2) what sources/claims-makers are quoted or referenced in the news item? Are environmental pressure groups or NGOs, for example, amongst the sources/claims-makers referenced?

What do the answers to these questions tell us about who or what drives/influences the news agenda on environmental issues?

To merely conclude that environmental journalism is authority-oriented is, however, also a gross simplification of a complex and varied situation. Thus, the influence of sources on media coverage may vary significantly from issue to issue: Hargreaves et al.'s (2004) analysis for example shows pressure groups to be much more prominent in media coverage of climate change than in media coverage of the measles–mumps–rubella (MMR) vaccine controversy or of cloning and genetic medical research. The degree of 'indexing' (Bennett, 1990) to national government policy also depends on the nature of event or issue being covered. Thus Shehata and Hopmann (2012) found the indexing to national government policy to be much less pronounced in relation to the major international climate change conferences (COPs) than in relation to day-to-day reporting on climate change. And Wozniak et al. (2016) showed a preference in news coverage for NGO-generated *visual* framing of climate change conferences, while the *textual* framing remained firmly aligned with government and official authority perspectives.

Almost invariably, environmental issues are the subject of considerable controversy, and rather than a single authoritative source, environmental journalists will tend to draw from 'both' sides or multiple sides, thus satisfying both the news value of 'controversy' and the journalistic values of balance and objectivity. This balancing of arguments has been shown to be prevalent in media reporting on many controversial environmental and science-related issues, but has been particularly prominently articulated and researched in relation to news coverage of climate change (Boykoff & Boykoff, 2004), but again it is important to recognise that here also the nature of journalistic reporting changes over time and varies considerably with the type of news medium (Hansen, 2016; Philo & Happer, 2013).

These more complex patterns are confirmed by the findings from a large and comprehensive study by Sachsman and his colleagues of environmental print and broadcast journalists in four major regions of the United States:

> State departments of environmental quality, local environmental groups, and local citizens active on the environment were among the most used groups. The results were a bit more varied at the bottom. Greenpeace was one of the least used sources in all four regions. The Chemical Manufacturers Association, the National Health and Safety Council, and the U.S. Food and Drug Administration were near the bottom in two regions, and two national agencies, the National Science Foundation and the U.S. Agency for Toxic Substances, were near the bottom in a single region.
> An examination of the use of all twenty-nine sources in each region showed an emphasis on local and state sources. Among national sources, only the Environmental Protection Agency was ranked between 2.0 (often) and 3.0 (sometimes) in all four regions.
> (Sachsman et al., 2006: 105)

The indications from research are that the power balance in the relationship between sources and journalists has shifted increasingly in favour of sources (Davis, 2013). Lewis et al. (2008) thus argue that:

> pressures on journalists to increase productivity, via substantive growths in the pagination of national newspapers across the last two decades, achieved with relatively static numbers of journalists [...] have prompted desk-bound journalists to develop an increasing reliance on pre-packaged sources of news deriving from the PR industry and news agencies.
> (1)

In their comprehensive study of UK print and broadcast media, Lewis and his colleagues found that in broadcast media 'the business world was nearly four times as likely as NGOs or pressure groups to "place" their PR material into news stories' (p.12); and that 'news, especially in print, is routinely recycled from elsewhere and yet the widespread use of other material is rarely attributed to its source' (p.18). They conclude:

> Our findings do, however, raise questions about the nature and sources of PR. As we have seen, it will favour those, notably business and government, best able to produce strong and effective PR material. It would however be unfair to blame journalists for relying on pre-packaged information. It is clear that most journalists operate under economic, institutional and organisational constraints which require them to draft and process too many stories for publication to be able to operate with the freedom and independence necessary to work effectively. What is clear from this study is that the quality and independence of the British news media has been significantly affected by its increasing reliance on public relations and news agency material; and for the worse!
>
> (Lewis et al., 2008: 18)

While the study by Lewis et al. does not look specifically at environmental news coverage, there are indications from elsewhere (e.g. Trench, 2009; Allan, 2006; Williams, 2015) that the exponential growth in online and Internet journalism, witnessed since the 1990s, has impacted particularly on science, health and environmental journalism (Trench, 2009: 175). This is complemented by a growing body of evidence on the increasing amount of source-generated pre-packaged material in general news reporting (Davis, 2013; Lloyd & Toogood, 2015). Economic pressures and organisational pressures have led to journalism that is increasingly desk-bound, which in turn has increased the scope for proactive news sources and news providers to 'subsidise' the work of news organisations and their journalists with ready-packaged and advantageously framed 'information', while at the same time depriving journalists of some of their most traditional networking and source-checking strategies based around 'face-to-face' interviews or contacts with sources (Williams, 2015).

Objectivity, balance, accuracy and the journalistic construction of expertise

It is not unusual for journalists and media to be made a scapegoat in relation to public debate involving controversial environmental or scientific claims-making, where different claims-making institutions and organisations are competing to promote their particular view. Indeed, a large body of media research, operating from a 'reflection-of-reality' and transmission of information perspective has focused on traditional questions about 'accuracy', 'objectivity' and 'bias/distortion' in media reporting of science and environmental issues (see Hansen, 2016, for a review of the changing uses of 'accuracy' in mediated communication about science and the environment).

A major problem in criticism of 'media accuracy' is the notion that the perceived inaccuracy is primarily a product of sloppy reporting, inadequate training of reporters, and actual media distortion. What is ignored in much of the criticism of media reporting of controversial issues is that the media in their reporting often simply reflect uncertainty, disagreement and controversy in the communities, scientific/industrial/political, which they report on. That pressure groups, industry, government, public authorities, and so on, deliberately package their claims with a view to enhancing and promoting their particular definitions of controversial issues should surprise no one. But that scientists and scientific institutions, far from fitting the image as independent establishers of scientific 'truth' and 'fact', are often embarked on similar projects, may be rather more contrary to the conventional wisdom which has guided research on 'accuracy' in media reporting. Yet, as Hilgartner (1990: 531) argues, 'a mountain of evidence shows that experts often simplify science with an eye toward persuading their audience to support their goals'.

Equally important for understanding how and why some issues become news and others do not, is to recognise that there is often competition and downright conflicts of interests between the various groups, institutions and agencies who make claims about the environment (Hilgartner & Bosk, 1988). Journalists have a difficult balancing act to perform amid a barrage of conflicting claims about environmental issues. A cornerstone of the professional ideology of news work is the need not to be, or not to be seen to be, simply the extended mouthpiece of any one camp or source. Journalists are generally very dismissive of anything which appears as thinly veiled promotional publicity, from whichever type of source (government, industry, corporations, science institutions, NGOs, etc.). But the task of performing the traditional journalistic role of investigative, balanced, critical and accurate reporting has become infinitely more challenging in a much changed media and communications landscape. As Williams (2015: 202) notes, this is a news environment where fewer environmental journalists are 'asked to do far more with no extra resources', with significant negative implications for the ability to execute traditional journalistic tasks, such as fact-checking stories for accuracy, exacerbating 'an already extant shift in the balance of power between reporters and their sources'.

Perhaps the most prominent articulation of accuracy concerns in the present century has so far been what has become widely known and recognised as the 'balance as bias' argument (Boykoff & Boykoff, 2004), most prominently articulated in relation to research on the public mediation of research on climate change. Much-cited research in this vein in the context of climate change are Boykoff and Boykoff (2004), Corbett and Durfee (2004) and Oreskes and Conway (2010), but the argument has been part of the accuracy debate rather longer (Hansen, 2016). The argument itself is admirably simple and elegant in its demonstration that media representations at odds with scientific consensus on an issue are a largely inadvertent result of the journalistic values of balance and objectivity. What is perhaps new to the long-standing research and debate regarding accuracy and objectivity in reporting is the growing body of evidence showing that traditional journalistic practices and values regarding balanced and objective reporting are being deliberately exploited by sources and claims-makers keen to manipulate the public communications environment to promote their particular ideological positions, through for example, the strategy of 'sowing doubt' about whether climate change is happening, whether it is caused by human activity, whether there is enough scientific evidence to take action, and so on (Oreskes & Conway, 2010; Miller & Dinan, 2015).

Journalists, who are of course generally not – nor could they reasonably be expected to be – experts on the science or social science behind the environmental issues and problems on which they report, deploy a number of key journalistic strategies for maintaining the legitimacy and credibility of their reporting. These strategies revolve around the core journalistic value or norm of 'objectivity', which involves, as noted above, abstaining from 'taking sides' and from 'advocating' the case of one claims-maker over another. Objective reporting must be – and more importantly, must be seen to be – balanced, accurate and based in 'facts' originating from credible sources (Dunwoody, 2014). The journalistic norms of objectivity and balance arose, as Dunwoody (2014: 33) notes, as 'surrogates for validity, that is, as ways of compensating for journalists' inability to determine whether their sources' assertions are true or not'.

While specialist journalists reporting on environment, science and health issues generally have considerable ('veteran' in Sachsman et al.'s terminology) experience of reporting on their fields, numerous studies have shown (e.g. Dunwoody, 1979; Friedman, 1986; Hansen, 1994; Stocking, 1999; Conrad, 1999) that they are acutely aware of the (scientific) uncertainties and controversy which often characterise debate about environmental issues. But, like journalists in general, and like other types of specialist journalists more specifically, they command an elaborate set of journalistic routines geared towards securing the credibility and 'objectivity' of their reporting. These include judging the credibility of their sources on the basis of such standard clues as qualification, age, seniority and institutional affiliation; using principally senior or top-ranking sources; using 'known' sources (that is, 'known' to the journalist, or to the research council, professional association or other source conduit that the journalist might use for tracking down relevant sources); they actively seek to cultivate a relationship of mutual trust with their sources, and in particular with a core of regular sources, to whom they turn – time permitting – and use as 'sounding boards' when dealing with new or 'unknown' sources (Hansen, 1994). The importance, to these specialist journalists, of trust and the cultivation of trust in the relationship with their main sources remains a core characteristic of journalistic values (Geller et al., 2005; Priest, 2015).

Frames and forums

While numerous studies (Dunwoody, 1979; Friedman, 1986 and 2004; Hansen, 1994; Stocking, 1999; Sachsman et al., 2006, etc.) of specialist correspondents covering the closely related journalistic beats of science, environment, health and medicine, agriculture, technology, and so on, have displayed a remarkable degree of consensus around the journalistic characteristics delineated above, these studies have generally had less to say about the importance of *framing* and *forums* in the journalistic construction of expertise and 'authoritative' coverage. The use of particular frames, settings (Ibarra & Kitsuse, 1993; see also Chapter 2) and forums for constructing or reinforcing the notion of expertise, credibility and authoritativeness is especially pronounced in television and other primarily 'image-based' media, which of course in the digital media age is a label that describes the majority of public communications media. Yet, while studies in other fields of news journalism (now-classic studies of media coverage of industrial conflict or of crime being cases in point – e.g.

Glasgow University Media Group, 1976; Hartmann, 1975; Hall, 1981) have commented extensively on the importance of framing, settings and forums for the construction of expertise and credibility, the analysis of how these dimensions are visually constructed in environmental news and journalism has only made significant progress in the present century (Hansen, 2018).

In their comprehensive analysis of television news and the visualisation of climate change, Lester and Cottle (2009) thus demonstrate how different key actors (politicians, scientists, environmental protesters, victims of climate change, etc.) are visually constructed in ways which associate very different degrees of authority, credibility and trust with these actors. Rebich-Hespanha et al. (2015: 512) likewise importantly note the different visual framing of ordinary people compared with authority figures. Ordinary people are depicted as 'suffering impacts of environmental conditions or engaging in efforts to mitigate or adapt', while authority figures are shown in active agency roles studying, reporting (scientists), or urging or opposing action (political figures and celebrities). As the authors conclude, this conveys very different visual messages about who are invested as authoritative 'agents of definition' for environmental issues and, on the other hand, ordinary people whose voices are marginalised.

Studies on the construction of scientific expertise in television documentary and factual programmes (e.g. Collins, 1987; Murrell, 1987; Hornig, 1990) have long demonstrated some of the key narrative conventions and visual props deployed in factual television programmes for conveying the authority of science and scientists. Likewise, it clearly makes a difference to the message which comes across regarding credibility, authority and expertise, if the argument of one side of an environmental controversy is represented by well-dressed, articulate and 'official-looking' government representatives interviewed in the hallowed halls of power, while the other side is represented by cold, wet and perhaps 'scruffy-looking' demonstrators gathered on a muddy verge somewhere near a major airport (see Exercise 4.2 below on framing and forums in news coverage of environmental protest).

In their comparative study of environmental news coverage in Britain and Denmark, Hansen and Linné (1994: 381), for example, found that not only do pressure group sources have an overall comparatively low profile, but when they '*do* appear as primary definers, they do so through the forum or news scenario of public demonstration or protest action rather than as legitimate or authoritative sources in their own right'. The construction of 'legitimate' expertise and authority is thus closely linked to questions about the settings, forums or arenas in which sources are 'placed' and filmed/photographed by journalists or with which sources are verbally 'associated' in journalistic print media reports.

There is clear evidence (Albaek et al., 2003) that journalists increasingly deploy expert and scientific sources in their reporting, and that journalism and media, like indeed other arenas in society, have been subject to an increasing 'scientisation' over the last half century or so (see also Bauer, 1998, on the 'medicalisation' of science reporting since the 1930s). Paradoxically – perhaps – the increasing journalistic deployment of expert sources parallels a trend of growing public distrust in 'authority' and in experts. As Boyce (2006) comments:

> Many researchers have observed the paradox of the decline in trust of experts and at the same time, the increasing use of expertise in Western society, '(w)e believe less and less in experts . . . (but) we use them more and more' (Limoges, 1993, p. 424).
>
> (890)

If increasing journalistic use of expert sources is perhaps in part a response to increasing public distrust in authority and experts, then it is also fair to assume that the journalistic task of 'constructing' reliable, credible and believable accounts may need augmenting in other ways as well, namely by way of more elaborate visual framings which can lend additional symbolic legitimacy and weight to the 'evidence' presented. Analyses of environmental journalism and environmental news reporting will thus benefit, as is already becoming clear from the emerging body of visual environmental communication research (Hansen, 2018), from moving well beyond text-focused analyses of how 'pro and anti' expert testimonies or sources are 'balanced' in media reporting to a multimodal analysis of how different expert testimonies are imbued with varying degrees of legitimacy and authority, or even in some cases positively undermined, by the setting, visualisation and other framings deployed by journalists.

EXERCISE 4.2 Framing and forums: the visual/verbal construction of environmental protest

Using the websites of major television news organisations such as CNN (cnn.com) or the BBC (bbc.co.uk), identify online news footage of a public protest or demonstration with an environmental component/

theme (e.g. fracking, airport expansion, genetic modification of agricultural crops, climate change).

What do the images show? – e.g. orderly, organised and peaceful protest or violent clashes between protesters and police?

How are the protesters/demonstrators portrayed? – e.g. does the camera focus on a 'cross-section' (age, dress, ethnic origin, etc.) of protesters or do close-up shots tend to pick out or focus in on 'extreme' or 'unusual' protesters?

Who is interviewed on screen? And where, i.e. against what backdrop?

How are interviewees framed visually and verbally? – i.e. how are they introduced or labelled? Where are they at the time of interview? What if anything goes on in the background as they are interviewed? How does the reporter's commentary link or juxtapose different interviews/interviewees? How does background noise and/or the reporter's commentary frame or inflect the authority/credibility or 'weight' of what the interviewees say?

The answers to these questions will give an indication of the highly constructed nature of visual news reports, alerting us to the fact that even seemingly straightforward video footage of a demonstration or protest action is essentially a selective construction resulting from a series of deliberate choices about what to focus on, what to juxtapose with what, how to narratively frame and 'give meaning' to that which is portrayed or reported. Ultimately, the answers to the above questions also enable us to determine the general stance and tone of the reporting, i.e. whether the protest is portrayed as warranted and legitimate, or alternatively as 'problematic', 'an unnecessary overreaction', 'a misuse of the right to demonstrate', etc.

Political and economic pressures: local/regional media reporting

Other pressures on the journalist, when deciding on what to cover and on how to frame the coverage, relate to the political and economic allegiances of individual media. Thus, there is a growing body of research on environmental journalism and communication to show the important linkages between media ownership, political and economic interests and the framing and stance of environmental reporting (Boykoff & Yulsman, 2013).

Political and economic pressures may include the actual or assumed (by the journalists) concerns of editors and media proprietors that some kinds of environmental coverage may antagonise powerful industries and businesses, whose advertising the media depend on. There is some indication that such pressures apply particularly in relation to media serving smaller regional or local communities.

Although most studies have focused on national prestige media, some studies have examined the coverage of environmental issues in local/regional media (Lahtinen & Vuorisalo, 2005; Wakefield & Elliott, 2003; Cottle, 1993 and 2000; Campbell, 1999; Gooch, 1996), highlighting some of the particular characteristics and constraints which govern local/regional media practices and distinguish them from national prestige media. Of particular interest here is the tendency for local/regional media to give more access to ordinary lay 'voices', as sources used in the definition and elaboration of environmental issues (Crawley, 2007; Cottle, 1993 and 2000), contrasting with the distinctive authority orientation of national news. However, the greater accessing of lay and oppositional voices is by no means a universal feature of local/regional media coverage of environmental issues, as other studies have confirmed a degree of authority orientation comparable to that usually found in national media (e.g. Corbett, 1998; Taylor et al., 2000).

Local/regional media have also, not surprisingly, been shown to focus on local rather than global problems, to celebrate a nostalgic view of nature and the countryside (Cottle, 2000), and to focus on problems relevant to or arising from the region served by such media. The latter, however, may be paralleled by a somewhat diminished scope for critical, investigative or adversary reporting, particularly where the 'causes' of environmental problems may be industries or institutions of major significance to the local economy (e.g. Dunwoody & Griffin, 1993). As in other fields of reporting, local/regional news media are conscious of and sensitised to the way in which local readership feelings may run high in relation to what might be construed as a criticism of local/regional institutions and businesses important to the local economy. Tichenor et al. (1980) and Donohue et al. (1989 and 1995) have shown how the scope for critical reporting and for accessing of a wider variety of voices/sources in the media relates closely to the size and degree of pluralism of local/regional communities. In more pluralist communities, the local media thus tend to be less authority-oriented, providing a platform for a wider range of viewpoints and positions, while in smaller and more homogeneous communities, the media tend to serve as a sentry not for the community as a whole but for the dominant groups of power and influence (Donohue et al., 1995).

As with other aspects of environmental news reporting, the dynamics of political and economic influences on local news and communication about the environment and environmental issues are changing in a globalised economy and in the digital and increasingly globally connected media landscape. Traditional concerns about alienating powerful employers in the local economy may be surpassed and overridden by a need for local media to voice local political disillusionment and alienation with regard to the political process, frustration with a perceived disenfranchisement with regard to decisions on controversial issues such as fracking, and local concerns about risks, pollution and safety in the immediate environment. Lester and Cottle (2015: 108) show how the dynamics of local environmental reporting are inevitably affected by the increasingly interconnected communications environment, involving 'simultaneously a politics of representation and politics of connectivity, with both necessarily involved in the scaling-up and scaling-down of local-global concerns'.

Event orientation, time and visualisation in the construction of environmental news

News is largely event-focused and event-driven, and it is this inherent event orientation which is an important factor in determining which environmental issues get news coverage and which don't. While some environmental issues are associated with spectacular and dramatic events – and thus eminently 'newsworthy' – many environmental issues are characterised by their relative invisibility. The thinning of the Earth's protective ozone layer is invisible to the naked eye; the detrimental effects of carbon dioxide on the Earth's atmosphere are not immediately observable; the link between pollution and rises in the incidence of respiratory diseases is not immediately obvious; the impact of chemicals in the environment may take years or decades to manifest themselves; likewise the impact of plastics/micro-plastics, and so on.

The invisibility and slow development of many environmental problems pose particular difficulties for their news construction and communication generally, and for their visual representation specifically (Hansen & Machin, 2013). In an early study of environmental news,

Schoenfeld et al. (1979) noted that one of the key challenges for news coverage of environmental issues was the mismatch between the demands of a rapid news cycle and the often long-term and slow-evolving pace of many environmental problems. But the challenge for news reporting extends significantly further than a mismatch of cycles: many environmental problems are just not that visible and the substances that cause them may themselves be invisible or innocuous-looking (Peeples, 2013). These are, as Hansen and Machin (2013) note, characteristics that make environmental problems and their visual representation more open – than, for example, their representation in text/language – to interpretation and indeed to ideological manipulation. Astute use of visual communication also potentially opens up for more effective claims-making, as shown by Wozniak et al. (2016) in an analysis of textual and visual news coverage of selected key climate change summits – the Conference of the Parties (COP): they show that while the textual discourse is dominated by political delegations and other traditional authority figures, the visual coverage is dominated by NGOs and environmental pressure groups, who are adept at exploiting the needs of news organisations and journalists for newsworthy visuals.

While some of the most prominent environmental issues in the media over the last few decades have been, intrinsically, neither particularly visible, nor well matched to the conventional timescales of news work, their prominence in the news is testimony perhaps that with the right kind of claims-making, visualisation, exploitation of journalistic and news values and exploitation of news routines, any potential environmental issue or problem can be packaged and constructed in ways that will attract the news media. Insight into the intrinsic event orientation of news and news work is thus strategically important for environmental claims-makers in that the ability to link campaign objectives and claims-making to events (e.g. upcoming national or international political, economic or cultural meetings or otherwise scheduled gatherings or negotiations) that predictably will attract news attention is a good guarantor of news coverage. Ungar (1992) uses the apt term or analogy of 'piggybacking' – 'environmental claims are most often honored when they can piggyback on dramatic real-world events' (Ungar, 1992: 483). However, the real-world events of course need not necessarily be 'dramatic' – although, in news value terms, drama certainly helps – but can simply be the kind of events taking place in those forums (e.g. parliament, the courts, international politics – and increasingly, in the digital media environment, social media) which by definition are attended to by journalists and news organisations.

Cultural resonances and frames in media coverage of environmental issues

If traditional approaches to the study of media and news coverage can throw some light on the relationship between the immediate actors involved in the production of such coverage, it is to the wider notion of *cultural resonances* that we must turn in order to account further for the very different careers and media presence enjoyed by different environmental problems.

Schudson (1989), Gamson (1988), and Hilgartner and Bosk (1988) have all argued for the importance of complementing traditional organisational perspectives on media coverage and news production with a wider view which takes into consideration how the 'cultural givens' of society both facilitate and delimit the elaboration and coverage of issues. In order to gain prominence in the public sphere, an issue has to be cast in terms which resonate with existing and widely held cultural concepts (Gamson & Modigliani, 1989).

Developing their public arenas model to explain the rise and fall of social problems, Hilgartner & Bosk (1988: 64) similarly note that selection principles of all institutional arenas, including the communications media, are 'influenced by widely shared cultural preoccupations and political biases. Certain problem definitions fit closely with broad cultural concerns, and they benefit from this fact in competition.'

Among the 'cultural givens' within which much media reporting on the environment is anchored are the beliefs in 'mastery over nature' (or 'nature as object', see Evernden, 1989) and in 'progress through science and technology'. Both of these may help explain why media discourse on the environment is – particularly in the early stages of reporting on new environmental issues – to a large extent a 'science' discourse drawing on scientists as the primary arbiters of right and wrong, true and false, real and imagined.

As Gamson and Modigliani (1989) maintain, for each cultural theme there is a counter-theme, and this principle is perhaps particularly evident in much environmental coverage with a heavy science component. The pro-science optimism which finds expression in themes and terms such as 'progress', 'development', 'industrialised', 'rational', and 'efficient' finds its counter-expression in numerous mediated forms, including perhaps most prominently, in popular films which stress the dangers of man's hubris and associated tinkering with nature. Indeed, a raft of studies across Europe as well as in the US have found, *inter alia*, the deep-seated cultural themes of 'Pandora's box', 'runaway technology/science' and 'tampering with nature' to be prominent – albeit to varying extents over time – in much news coverage of developments in genetic modification and biotechnology (Bauer et al., 1999; Nisbet & Lewenstein, 2002; Ten Eyck & Williment, 2003; Hansen, 2006) as well as in news reporting on the environment in general (Nisbet & Newman, 2015).

The notion of 'cultural resonance', as a contribution to understanding why some issues gain currency in public and media debate more easily than others, can be extended further to include questions about the ease with which some issues link into powerful, historically established, symbolic imagery. This marks out nuclear power issues and a host of genetic or biotechnology-related issues with environmental dimensions from some other environmental issues (e.g. ozone depletion, climate change, deforestation, species extinction, etc.). Patterson (1989) for example, drawing on psychological and risk studies, notes the deep-seated public fears ('Atom-Angst') associated with anything nuclear. Others have pointed to the powerful and deep-seated images of mass destruction associated with the use of nuclear bombs in the Second World War. Spencer Weart (1988), in an impressive historical analysis of the origins and inflections of 'nuclear images', traces the fear and angst-ridden imagery much further back to the discovery of radiation, to 'alchemistic' connotations associated with the new science of atoms and nuclear radiation, and to the key theme of 'mutation/transmutation' – themes which are all equally central to the rapid developments witnessed in biotechnology and genetic modification in the last few decades.

Metaphorical anchoring and linking of arguments to widely available cultural values and images give such images more potency in the public mind and in public debate, and in turn they serve to *frame* public perception and debate about their subject matter. Patterson's (1989) analysis of American and other Western news coverage of the Chernobyl nuclear accident similarly notes the important ideological implications of the ways in which the accident was framed. Principally, the accident was framed in terms of the dangerous technological incompetence of the Soviet Union, a frame which placed the blame firmly on the particular design of Soviet nuclear technology and the (in-)competence of its engineers, operatives and bureaucrats, while specifically diverting attention away from Western nuclear technology ('it could not happen here') or suggestions that civil nuclear energy generation is inherently dangerous and prone to 'normal' accidents.

Much framing and the activation – through vocabulary and metaphor choice – of particular cultural resonances start with the claims-makers who make claims about environmental issues. But, as we have seen in the highly selective use of sources, the media play a key *gatekeeping* role through their control over the *selection* of sources, information and arguments. In addition, they do further 'ideological work' by adding their own discursive spin or framing to the issues on which they report:

> the media can impose its own meaning frames and symbols to a given event. By calling a policy 'controversial', by highlighting a dispute, by suggesting a benefit and excluding risk information, the media can legitimate positions and project images with considerable power [...] Lee (1989) [...] demonstrated the significant role played by opposing sides in framing the debate [about seal hunting] in strong moral terms. More interesting, however, was the additional role played by the media in injecting its own voice in the debate. The opponents to the seal hunt led by Greenpeace claimed the hunt was a 'slaughter' of 'endangered baby seals'; the hunters, on the other hand, succeeded in gaining support for their cause by presenting the attempts to stop the hunt as 'cultural genocide'. Lee further demonstrated that the largest amount of moral keywords presented were not from the competing claims-makers but in the reporters' own voices. He concluded thus: 'the moral reality of the seal hunt is communicated in the apparently objective voice of the news reporter rather than [...] in the voices of the moral contestants, a profoundly significant comment on the power of the newspaper to construct moral social reality.'
>
> (Einsiedel & Coughlan, 1993: 136–137)

In their analysis of nuclear discourse, Gamson & Modigliani (1989) point out that different, and often competing, frames may coexist simultaneously in different forums of meaning-making (e.g. policy-making forum, the media, public opinion, etc.). Smith (1992: 109) examined this phenomenon in relation to media coverage of the Exxon Valdez oil spill in Prince William Sound, Alaska, and found that the event was framed variously as 'environmental catastrophe' and 'as a fable about a drunken sea captain and a mighty oil company that couldn't clean up after itself'.

Daley and O'Neill (1991: 42) point perhaps even more succinctly to the ideological implications (what causes, who to blame, what appropriate remedial action) of the various frames or narratives used in media coverage of the Exxon Valdez oil spill. From their analysis of media coverage, they conclude that 'mainstream narratives naturalise and individualise the spill, turning ordinary members of the public into victims, while minority narratives offer a competing conception of nature'. They identify three mainstream frames or narratives (a disaster narrative, a crime narrative and an environmental narrative), and one marginalised minority frame (a subsistence narrative) in the coverage:

> The disaster narrative naturalized the spill, effectively withdrawing from discursive consideration both the marine transport system and the prospective pursuit of alternative energy sources. The disaster narrative overtly moved the discourse away from the political arena and into the politically inaccessible realm of technological inevitability [...] The disaster narrative also drew upon a few widely available cultural metaphors – Faustian bargains, Frankensteinian nightmares – to transform a specific social and historical event into a tale of natural or technological inevitability [...] The disaster narrative and journalistic practice in general facilitate the categorisation of accidents as abnormal, thus deflecting attention away from production-driven systemic problems.
>
> (Daley & O'Neill, 1991: 53)

The *crime* narrative framed the event in terms of individual failure and incompetence, focusing attention on the drink problem and personality of the captain, and on legal proceedings against him. The *environmental* narrative focused on environmental spokespersons contesting the statements and practices of industrial spokespersons and government – but rather than highlighting the threat which the oil spill posed to the livelihood of Native Alaskans (the marginalised *subsistence* narrative), this was a frame which drew on traditional binary oppositions of messy, high-tech, industrial exploitation versus a (largely romanticised) view of nature as a place of tranquil and pristine beauty.

> This narrative frequently offered visually compelling photos of soiled birds or cuddly sea otters, domesticated before our eyes and for our eyes. They were offered as pathetic human interest stories, not as animals who exist in their own right in fragile marine ecosystems.
>
> The spectator role scripted for the public by the environmental narrative is at odds with Native Alaskans' subsistence relationship with nature. Because Native Alaskans see themselves as a part of this system, they see no need to anthropomorphize animals.
>
> (Daley & O'Neill, 1991: 54)

The extent to which the images and frames used in environmental reporting resonate with cultural themes or with emotions can be as important in effective environmental communication as the sheer frequency and repetition of images or messages. In their study of media coverage and visual agenda-setting regarding the BP Deepwater Horizon oil spill disaster in the Gulf of Mexico in 2010, Miller and LaPoe (2016) show that – as expected – the most prominent images are also those best remembered by the public. But their study further shows that particularly emotional or resonant images, such as images of oil-soaked animals, become the most memorable even when they are not the most frequent or numerically prominent.

More generally, and building on Gamson and Modigliani's (1989) classic analysis of media discourse and public opinion on nuclear power, studies of mediated environmental communication have demonstrated how culturally resonant interpretative packages (*progress*, *economic prospect*, *ethical/moral*, *Pandora's box/runaway science*, *nature/nurture*, *public accountability*, etc.) are drawn upon by sources and journalists alike, and persist over time, in the communication of environmental issues (Nisbet & Newman, 2015). Such frames or packages are strategically deployed and manipulated in public environmental debate, with significant potential implications for both the nature of media representation and the mobilisation of public understanding of controversial issues such as nuclear energy, human and agricultural biotechnology, climate change, and so on.

Not all environmental issues or problems engage with, or 'benefit' from, a culturally deep-seated imagery of the same symbolic richness as genetics/biotechnology or as 'nuclear/radiation-related issues', and they are disadvantaged by this in competition for elaboration in media and other meaning-creating forums. But it is precisely the extent to which they can be anchored in and made to activate existing chains of cultural meaning which helps determine whether they become successful claims, gain prominence in media coverage, achieve legitimacy in public arenas, and influence social and political action.

Conclusions

News coverage of the environment and of environmental issues is the result of complex processes of construction. Research on the 'sociology of news' since the 1950s has indicated how factors such as news values, organisational structures and arrangements in media organisations, economic and political pressures on news organisations, the professional values and working practices of journalists, and so on, all influence to varying degrees what is reported and how it is reported.

The organisational arrangement of having an 'environment beat', staffed by specialist 'environment reporters', has been shown to considerably influence the amount and nature of coverage of environmental issues. Studies of specialist reporters in the fields of science, environment and health reporting have added further to our understanding of how and why these areas are reported. They have pointed to the complex strategies that journalists deploy in order to deal with the pervasive 'uncertainties' characteristic of issues in these areas, including by 'balancing' opposing sources and by cultivating relationships of trust with sources, by drawing on recognised, 'credible' and 'authoritative' sources.

Recent studies have shown how technological advances and increasing economic pressures on news organisations combine to change the nature of newsgathering and news work generally, and how this has impacted particularly on specialist areas of reporting such as science and the environment. The evidence discussed in this chapter thus points towards environmental journalism which is under increasing pressure, resulting – through a combination of technological advances, economic pressures as well as a changing political and 'publicity-driven' climate – in a shifting balance of power between journalists and their sources, moving in the direction of sources. Gans' observation in the 1970s that the relationship between journalists and sources is like a dance, but while 'it takes two to tango [and] either sources or journalists can lead, (…) more often than not, sources do the leading' (2004 [1979]: 116) thus seems to be truer than ever before.

Studies of media organisations and different types of media (broadcast versus print media, national versus regional/local media, 'quality/elite' versus 'popular' media) have provided further evidence of the complex factors which impinge on what is and can be said about environmental issues; particularly interesting – although as yet insufficiently researched and understood – are perhaps the very different obligations, responsibilities or pressures which characterise regional/local media compared with large national and international media, particularly when it comes to the balancing of economic, employment and environmental considerations with particular relevance to regional/local communities.

The final part of the chapter aimed to go beyond the focus on organisational characteristics and arrangements of news media and the professional values and practices of journalists to indicate how wider cultural factors also impinge on the nature of environmental news. Put simply, certain environmental issues or stories 'resonate' more easily with common cultural narratives or preoccupations than others; certain ways of 'saying things' (choice of words, metaphors, images) or reporting on issues/problems sound more familiar and perhaps therefore more plausible and trustworthy to media publics than others. In the next chapter, we shall pursue these 'cultural packages' and cultural resonances further by tracing their origins in and perpetuation through media and popular culture more generally, that is, beyond the present chapter's particular focus on news and news media.

Further reading

Friedman, S. (2015). The changing face of environmental journalism in the United States. In A. Hansen & R. Cox (Eds.), *The Routledge Handbook of Environment and Communication* (pp. 144–157). London and New York: Routledge.

Sachsman, D. B., & Valenti, J. M. (2015). Environmental reporters. In A. Hansen & R. Cox (Eds.), *The Routledge Handbook of Environment and Communication* (pp. 158–167). London and New York: Routledge.

Williams, A. (2015). Environmental news journalism, public relations and news sources. In A. Hansen & R. Cox (Eds.), *The Routledge Handbook of Environment and Communication* (pp. 197–205). London and New York: Routledge.

Anderson, A. (2015). News organisation(s) and the production of environmental news. In A. Hansen & R. Cox (Eds.), *The Routledge Handbook of Environment and Communication* (pp. 176–185). London and New York: Routledge.

Dunwoody, S. (2014). Science journalism: prospects in the digital age. In M. Bucchi & B. Trench (Eds.), *Handbook of Public Communication of Science and Technology* (2nd ed., pp. 27–39). London: Routledge.

5 Popular culture, nature and environmental issues

THIS CHAPTER:

- Discusses the notions of scripts, cultural packages, interpretive packages and cultural resonance, and explores their significance for understanding the nature and potential 'power' of popular media representations of nature and the environment.
- Traces the particular interpretive packages which have been – and continue to be – prominent in media and popular culture constructions of the environment, drawing particularly on research on representations of nuclear power, biotechnology and climate change.
- Explores some of the core frames and cultural assumptions which are in play in media and public sphere discussions about the environment, nature and environmental issues.
- Looks beyond the conventional focus on *news coverage* of the environment to explore how nature and the environment are constructed ideologically in other media genres, including television entertainment programming and more particularly in wildlife film and television nature programmes.
- Explores the significance of lexis or word choice in media constructions of environmental issues, and the similarly significant contribution of narrative analysis to uncovering the deeper ideological values communicated through wildlife film and nature programming.
- Discusses the historical changes in narrative and stylistic formats, and explores the relationship between socio-historical circumstances and media constructions of nature and the environment.

Introduction

Images of nature and the environment extend far beyond news media reporting. As indicated at the end of the previous chapter, the success of environmental claims in the news media will often depend on the extent to which such claims engage or resonate with deeper cultural images, beliefs and perceptions. This chapter therefore focuses on research into the deeper-lying cultural and interpretive packages which can be seen to inform media and public images of nature, the environment and environmental issues. Drawing on studies of a range of popular media genres, the chapter will show how deep-seated cultural narratives have reflected, and in turn shaped, particular ideological interpretations of nature and the environment, including changing dominant interpretations of the environment as either an object of control and exploitation or as something to be protected.

Much of communications research focusing on the media coverage and media reporting of the environment and environmental issues has focused mainly on news coverage (on the – often implicit – understanding that news coverage is the main influence on everyday political opinions, public understanding, etc. of socially constructed issues like environmental controversy). Likewise, such analyses have often been concerned with traditional questions about the amount of coverage, accuracy and balance/bias in coverage, 'primary definers' (i.e. the key sources quoted or referred to in news coverage) and major thematic emphases of coverage. These are all important aspects of any analysis of news coverage. They are also categories that are generally well-grounded in relevant theoretical frameworks for understanding the nature and social functions of news. Thus, an emphasis on balance and bias in news coverage relates to key concerns in theories about the role of news media and their watchdog role in society, as well as to core ideas about the professional practices and the required – by professional ideology – objectivity and impartiality in news reporting; the emphasis on 'primary definers' relates closely to social constructionist concerns about claims-making as well as to wider – often Marx and/or Gramsci-inspired – concerns about hegemony, inequality and the maintenance of power structures in society. But by focusing on these – more obvious and conspicuous – dimensions of news coverage, the deeper-lying and perhaps taken-for-granted assumptions, myths and ideologies which form both the basis and contexts for 'what is or can be said' about certain problems or issues have been less in focus and less foregrounded in analyses of media coverage of environmental issues and problems.

The rise since the early 1990s in the application of framing theory has provided a major impetus for a refocusing of media and communication research on these deeper-lying structures and messages, and this has helped in particular in throwing light on such key questions as why it is that some types of claims about environmental problems are or become much more successful in the public sphere than others. As we saw in Chapter 2, the types of rhetoric deployed (e.g. rhetoric of loss, rhetoric of calamity, rhetoric of unreason – see the discussion of Ibarra and Kitsuse, 1993, in Chapter 2), simple lexical choice, and the types of metaphors used reference, invoke or signal deeper-lying ideological 'scripts' or cultural narratives, that articulate – as ideological clusters – a particular perspective, world view, assumption or understanding.

In his excellent analysis of the deeper historical and cultural images which inform much of public debate about genetic modification and biotechnology, Turney (1998) notes the power of such scripts:

> Once a script has been laid down, a single cue can evoke an entire story, as an interpretive frame or context for what is being discussed. In this sense, the *Frankenstein* script has become one of the most important in our culture's discussion of science and technology. To activate it, all you need is the word: *Frankenstein*.
>
> (6)

As argued in the previous chapter, some environmental problem formulations 'resonate' more easily or better with deep-seated cultural assumptions and anxieties than others, and they benefit from this in their public sphere careers. Likewise, some ways of making claims about the environment 'resonate' or 'fit' better with established or conventional journalistic and news 'templates' than others, and consequently they stand a better chance of, first, gaining media attention, and second, of gaining favourable and legitimating media coverage.

The key objectives of this chapter are twofold: 1) to explore further some of the core frames and cultural assumptions which are in play in media and public sphere discussions about the environment, nature and environmental issues; and 2) to look beyond the conventional focus on news coverage of the environment, and to begin to show that these frames and cultural assumptions permeate throughout – and are in some respects the bedrock of – a broad range of different media genres and types of content. In this chapter we shall look particularly at the interpretive packages and frames furnished by the science fiction genre (literature and film), wildlife/nature films and popular television nature programming. In the next chapter we shall focus more specifically on the genre of advertising and the articulation of environment and nature in this particular persuasive genre.

Cultural packages, assumptions and frames in media reporting on the environment

Cultural packages, scripts, schema or narratives can be regarded as a 'reservoir', indeed they could be seen as the reservoir of world views which helps us understand how things work and what counts as appropriate or acceptable within any given culture. The notion of a reservoir of world views or perspectives, readily available to members of a particular culture, is akin to another sociological metaphor, namely Ann Swidler's notion of culture 'as a "tool kit" of symbols, stories, rituals, and worldviews, which people may use in varying configurations to solve different kinds of problems' (1986: 273). But while the reservoir or toolkit may be *always available*, it is also the case of course that not all parts of it are activated or drawn upon equally often (just as one – to continue the analogy – would rarely have all the tools out of the toolbag all at once, nor indeed would all of the tools be called upon equally frequently).

Moreover, the meaning, relevance and usefulness of particular clusters within the cultural reservoir change – or are deliberately changed – over time. As Gamson and Modigliani (1989: 2) put it, 'Packages ebb and flow in prominence and are constantly revised and updated to accommodate new events.' Changes in packages and their relative prominence in the public sphere depend on historical, social and political factors, and of course also on the more media-specific factors to do with the organisational, professional, technological and economic dimensions discussed in the previous two chapters.

In their now classic and oft-quoted article from 1989, Gamson and Modigliani offer an exemplary analysis and map of the *cultural packages* which have served as interpretive frameworks for media and public discussion about nuclear issues and nuclear power since 1945. As indicated in Chapter 2, Gamson and Modigliani offer a useful approach to frame analysis, and as we shall see in Chapter 7, they also offer a particularly helpful way of conceptualising the relationship between media coverage and public opinion.

Here, however, I wish to focus on their core proposition that analysis of media and public debate needs to move beyond the quantitative measurement – characteristic of much communication analysis of media coverage of nuclear power and related concerns about the media's (negative) influence on public opinion about nuclear power – of whether media coverage is positive or negative, for or against, nuclear power. Analysis needs to take into account the broader scripts or interpretive packages, which make up the reservoir of scripts, storylines, idea clusters, core propositions, and so on from which both media and the public draw to make sense of issues, in this case of nuclear power. As Gamson and Modigliani (1989) argue:

> [E]very policy issue [...] has a culture. There is an ongoing discourse that evolves and changes over time, providing interpretations and meanings for relevant events. An archivist might catalog the metaphors, catchphrases, visual images, moral appeals, and other symbolic devices that characterize this discourse. The catalog would be organized, of course, since the elements are clustered; we encounter them not as individual items but as interpretive *packages*.
>
> (1–2)

Packages, in Gamson and Modigliani's argument, are not intrinsically about positions for or against an issue; rather they provide the building blocks *and* a guide structure or frame for viewing a particular issue. They thus, for example, identify the *progress package* in the nuclear power discourse as a package that 'frames the nuclear power issue in terms of society's commitment to technological development and economic growth' (Gamson & Modigliani, 1989: 4). But not all packages 'speak to' or resonate equally powerfully with the underlying culture:

> Certain packages have a natural advantage because their ideas and language resonate with larger cultural themes. Resonances increase the appeal of a package; they make it appear natural and familiar. Those who respond to the larger cultural theme will find it easier to respond to a package with the same sonorities. Snow and Benford (1988) make a similar point in discussing the 'narrative fidelity' of a frame. Some frames 'resonate with cultural narrations, that is, with the stories, myths, and folk tales that are part and parcel of one's cultural heritage.'
>
> (Gamson & Modigliani, 1989: 5)

The progress package thus benefits from its resonance with deeper American values, in many respects overlapping with the Protestant ethic (Weber, 1930) of many European countries. It also resonates with a broader utilitarian enlightenment perspective emphasising efficiency, technological innovation, a frugal, rational and logical approach, science and – perhaps most importantly of all in the context of environmental issues – mastery over nature as the key to progress.

Focusing their analysis on the public discourse on nuclear power from 1945 onwards, Gamson and Modigliani (1989) identify the following seven packages, which help explain the general framing and emphasis in public communication about nuclear power: *progress*; *energy independence*; *devil's bargain*; *runaway*; *soft paths*; *public accountability*; and *not cost-effective*. Inevitably, some of these categories are particularly apt in relation to public debate about nuclear energy in the historical period covered by Gamson and Modigliani's analysis. But the main proposition, that public communication and debate draws from a basic reservoir of cultural packages and frames, and that these are key to understanding the dynamics of (successful and unsuccessful) claims-making in the public sphere, has productively informed much environmental communication research from the 1990s onwards. In their succinct survey of this line of inquiry, Nisbet and Newman (2015: 328) provide a more generalised set of packages or frames which have emerged from studies across a range of issues, countries and time periods since the 1990s:

- Social progress: At stake is improving quality of life, or finding solutions to problems. An alternative interpretation is progress defined as living in harmony with nature instead of mastery, 'sustainability,' 'balance,' and 'quality of life';
- Economic development/competitiveness: At issue is economic growth and investment, market benefits or risks; protecting local, national, or global competitiveness;
- Morality/ethics: The issue is fundamentally a matter of right or wrong; respecting or crossing religious, ethical or 'natural' limits, thresholds, or boundaries; and/or working towards justice for those who have been harmed;
- Scientific/technical uncertainty: The issue or decision is a matter of expert understanding; what is known versus unknown; arguments either invoke or challenge expert consensus, call on the authority of 'sound science,' falsifiability, or peer-review to establish criteria for decisions;

- Pandora's box/runaway science/fatalism: This is a call for precaution in face of possible impacts or catastrophe. It defines problem or technology as out-of-control, a Frankenstein's monster, or as fatalistic, e.g., action is futile, the train has left the station, the path is chosen, no turning back;
- Public accountability/governance: Is a decision or action in the public interest or serving private interests? Emphasis on fairness, transparency, ownership, control including responsible use or abuse of expertise in decision-making, e.g. 'politicisation';
- Middle way/alternative path: An issue or decision is about finding a possible compromise position, or a third way between conflicting/polarised views or options;
- Conflict/strategy: At stake is a broader power game among elites; emphasising who's ahead or behind in winning debate, in public opinion polling or political spending. Emphasis is on the battle of personalities; or groups; the tactics and strategies involved and how they will 'play politically' (usually journalist-driven interpretation).

As argued in previous chapters (see particularly Chapters 2 and 3), a key to understanding the dynamics of the claims-making process is the recognition that for every claim there's a counterclaim (or in Ibarra & Kitsuse's [1993] wording, a 'counter-rhetoric'). Gamson and Modigliani similarly make this point, and do so in a way which succinctly articulates one of the key concerns of this chapter, namely the focus on deeper-lying cultural scripts which can be evoked and drawn upon in the public discourse on environmental issues. In relation to the technological 'progress' values delineated above, they thus identify the equally deep-seated countervalues in western culture, namely the emphasis on harmony with and protection of nature rather than exploitation of it and control over it, the valuing of 'small-scale' harmony and quality rather than intrusive maximisation, efficiency and quantity. And if the progress theme is prominent – as Gamson and Modigliani demonstrate – in media and public discourse, the countertheme finds equally powerful expressions in popular culture: 'Much of popular culture features the countertheme: Chaplin's *Modern Times,* Huxley's *Brave New World,* and Kubrick's *2001* and countless other films about technology gone mad and out of control, a Frankenstein's monster about to turn on its creator' (Gamson & Modigliani, 1989: 6).

In this context, it is worth noting that the traditional 'progress package' which is closely associated with economic and technological progress – the idea that social progress is driven by scientific and technological advances – is, in Nisbet and Newman's articulation, rather more flexible and encompassing. It is also worth noting that studies of popular culture representations, including in film, in the present century have shown some of the countertheme packages – perhaps particularly the Pandora's box/runaway science/fatalism package – to be prominent in popular film and entertainment fare representations of everything from climate change to biotechnology (Svoboda, 2016).

Spencer Weart (1988 and 2012), in his pioneering archaeology of nuclear images in popular culture, shows one of the core scripts in relation to public images and discourse on nuclear power to revolve around deep-seated public fear about the interference with nature or the natural order in ways which are both unpredictable and potentially highly devastating. Weart argues that public narratives of the mid-20th century about nuclear bombs polluting fish, causing birth defects or influencing the weather system all amounted to saying that nuclear energy 'violated the order of nature' (Weart, 1988: 187–188):

> This idea was bound up with one of the strongest of primitive themes: contamination. In most human cultures the violation of nature, and forbidden acts or things in general, have been directly identified with contamination. According to the anthropology theorist Mary Douglas, whatever is 'out of place', whatever goes against the supposed natural order, is called polluting.
>
> (188)

Weart is discussing nuclear technology, but, as Turney (1998) and others have noted, the fundamental fear that scientific progress and science's interference with nature or with God's creation may spiral 'out of control' and produce dangerous outcomes that are both 'wrong' (an ethical discourse) and unpredictable is equally central to much of the history of genetic manipulation and biotechnology, and indeed across other fields. As with public fears about nuclear power, public fears about genetic manipulation draw from a cultural reservoir, originally expressed in literature and then in wider popular culture products, not least film. Mary Shelley's *Frankenstein*, H.G. Wells' *The Island of Dr Moreau* and Aldous Huxley's *Brave New World* – and popular films based on these novels – are perhaps the most prominent expressions and sources of such fundamental fears. Weart (2012) points to how public and popular images draw on culturally deep-seated ideas about contamination:

> Most important was the fact that radiation could cause genetic defects. This resonated with old and widespread ideas about contamination. Traditionally, defective babies were a punishment for pollution in the broadest sense, violations such as eating forbidden food or breaking a sexual taboo. [...]
>
> The association between nuclear radiation and pollution was strengthened by the fact that radiation can cause cancer. By the 1950s the word 'cancer' had come to stand for any kind of insidious and dreadful corruption. Demagogues labelled Communists, prostitutes, bureaucrats, or any other despised group as a social 'cancer' [...].
>
> Nobody set out deliberately to make any of these associations; they came to many people at once. As early as 1950, liberal newspaper and radio commentators had exclaimed that hydrogen bombs, wrongfully exploiting the 'inner secrets' of creation, would be 'a menace to the order of nature'.
>
> (101)

As in the public discourse on nuclear power, so too has the increasingly prominent media and public discourse on genetic modification and biotechnology been characterised by broadly two opposite discourses, each with deep cultural resonances: one, celebratory and enthusiastic about the immense potential of scientific and technological progress; and the other concerned and fearful about the dangers and potential for 'out-of-control' damage inherent to scientific and technological development.

Longitudinal analyses of media coverage of biotechnology and genetics both in the UK and the United States (Durant et al., 1996; Bauer, 2002; Bauer et al., 1999; Nisbet & Lewenstein, 2002; Nisbet & Huge, 2006; Listerman, 2010) show how the polarisation of public discourse on genetics and biotechnology became particularly pronounced in media coverage from the latter half of the 1990s onwards. Bauer (2002) and Nisbet and Lewenstein (2002) thus point to a significant change in the overall 'symbolic environment of biotechnology' in the 1990s in the form of a deepening polarisation between 'desirable' biomedical research/applications and 'undesirable' agri-food biotechnology, although it needs to be noted that there are deep cultural differences between Europe and North America in terms of relative prominence of these discourses.

Drawing closely on Gamson and Modigliani's notion of framing and their typology of cultural packages, communication research in Europe, Australia and the US (e.g. Bauer et al., 1999; Petersen, 2001; Nisbet & Lewenstein, 2002; Ten Eyck & Williment, 2003) has mapped key trends in media coverage of the new genetics and biotechnology, including the rapid advances in genetic manipulation of both plants and animals. It is in the nature of cultural packages that they articulate general ideas and principles rather than issue-specific characteristics. It is thus not surprising that many of the same cultural packages identified in the nuclear power discourse are equally prevalent in the media and public discourse on genetic engineering/biotechnology, and indeed on climate change (Nisbet & Newman, 2015). Thus, in public discourse on biotechnology, the 'progress' package (celebrating the rapid advances, breakthroughs and developments in genetic research and science) features prominently, as does the 'economic prospect' package. But also present, and in some cases increasingly so as the coverage moves into the more recent period, are the 'fear and concern'-related packages such as 'nature/nurture', 'Pandora's box' and 'runaway technology/science'.

Turney (1998), Huxford (2000) and others have persuasively argued that environment and science correspondents as well as journalists generally rely heavily on readily available cultural scripts and frames, particularly when reporting on new and unfamiliar developments in science and environmental issues. Likewise, much communication research – including environmental communication research (Väliverronen & Hellsten, 2002; Nerlich, 2003, 2010; Renzi et al., 2016; Healy & Williams, 2017) – has pointed to the central significance of metaphor in news reporting (see also the discussion in Box 5.1). This is not surprising, as the very purpose of metaphor is of course to make the unfamiliar familiar, or in other words, to help us understand that which we don't know by explaining it with reference to what we do know and are familiar with. Invoking familiar scripts, interpretive packages, schema or metaphors serves the purpose of transferring ('transfer' is the literal meaning of the Greek root of 'metaphor': *metapherein*) meaning from what we know to what we don't, as yet, know. Importantly, however, the process is not simply a matter of conveying information or establishing comprehension, but also significantly carries with it a particular 'view' or 'perspective', a particular way of framing the unknown, and with these, a potential attitude or stance on things. For journalists, invoking particular frames or scripts is often 'more to do with cueing certain cultural fears' (Huxford, 2000: 192), as embodied in popular culture, than with providing an understanding of environmental or scientific developments or phenomena.

But while our cultural reservoir is replete with both utopian and dystopian perspectives on nature, the environment and science, it is often the apocalyptic and dystopian (Svoboda, 2016), rather than the optimistic and utopian, scripts from literature, film and other popular culture that are invoked and drawn upon in media and public discourse on environmental and scientific developments, although considerable variations appear across different genres (Moore, 2017). This is particularly the case, where – as with nuclear power, biotechnology, and climate change – genuine fears, concerns and worries are fuelled by (perceived or real) scientific uncertainty or controversy. Like Turney (1998), Huxford (2000) thus for example demonstrates the prominence of particular science fiction frames or scripts in media reporting on cloning:

> The use of the science fiction frame was extremely common in this coverage, both in Britain and in the United States. In a sample of 204 articles in which cloning was the principle focus, ninety-two (46 per cent) carried such associations. Of these, the tone of the vast majority was negative, raising fears of the future use of the technology.
>
> (191)

Huxford's analysis shows the three most prominently referenced science fiction works to be Aldous Huxley's *Brave New World* (mentioned in 32 of 92 articles referencing science fiction), Mary Shelley's *Frankenstein* (referenced in 21 of 92 articles) and *The Boys From Brazil* (13 of 92 articles). The analysis shows, argues Huxford (2000):

> the way that science fiction frames, employed by the media during the clone coverage, brought with them largely negative narratives that cued a series of oppositional couplings: religion versus science, man versus institution, the individual versus society, high culture versus low culture, past versus future. In each case the former was privileged—but shown to be threatened through cloning—by the latter.
>
> (197)

The largely negative scripts identified in Huxford's analysis as readily used frames for media coverage of cloning interestingly bear many similarities with some of the dominant interpretive packages identified by Gamson and Modigliani in relation to popular culture narratives about nuclear power. Like the frames informing popular culture narratives about the new genetics, those concerned with nuclear power can similarly be divided broadly into the dichotomy of dystopian and utopian scripts/frames or narratives. But as shown in Gamson and Modigliani's analysis, the relative prominence of different interpretive packages varies across genres of media and popular culture output.

This is further confirmed by Podeschi's (2002) analysis of environmental discourse in science fiction films from the latter half of the 20th century. Podeschi thus demonstrates a marked change in this particular genre's nuclear discourse, from scary foreground narratives – often metaphorically transcoding and expressing the communist and cold war fears of the time – in the 1950s and 1960s, to a *naturalisation* and backgrounding of nuclear power as an innocuous and abundant power source for the futuristic societies commonly depicted in science fiction film. As Podeschi notes, such portrayal is at odds with both the rise of public resistance to nuclear power evidenced during the 1970s and 1980s, with the flood of news coverage brought on by the high-profile nuclear accidents at Three Mile Island (1979) and Chernobyl (1986), and with non-science-fiction film discourses on nuclear technology:

> Unlike the film *The China Syndrome* (1979) which transcodes fears about nuclear power, science fiction films [after 1970] are comfortable with the technology.
>
> Concern about nuclear weapons and concern about technology that unites with humanity are the strongest threads of resistance in the entire sample. Taken together, they cover the entire period, one largely prior to 1970 and one largely after 1970. Combined with the naturalisation of nuclear power that is also evident after 1970, these films may be articulating a shift in societal anxiety, from the tangible fear of radiation and nuclear war to explorations of futures in which computer technology is central and powerful in our lives.
>
> (Podeschi, 2002: 288)

EXERCISE 5.1 Interpretive packages in the current nuclear power and climate change debate

Media images and public perceptions regarding nuclear power have a turbulent history, characterised by ups and downs in both levels of interest and in 'climate of opinion' surrounding this technology. Following comparatively low levels of media and public interest in nuclear power during the 1990s, the start of the new millennium witnessed renewed interest:

> At the start of 2000, however, new focusing events began to shift the interpretative packages and mental categories applied to nuclear energy. In 2001, in reaction to rising energy costs and rolling blackouts in California, the George W. Bush administration launched a communication campaign to promote nuclear power as a *middle way path* to energy independence. The terrorist attacks of September 11, 2001, dampened the viability of this frame package, as experts and media reports focused on nuclear power plants as potential terrorist targets. But since 2004, as energy prices have climbed and as U.S. dependence on overseas oil has been defined by political leaders as a major national security issue, a renewed emphasis on the energy independence interpretation has surfaced.[...]
>
> The effort by the second Bush administration and the nuclear energy industry to reframe the relevance of nuclear energy has been complemented by an attempt to similarly sell nuclear energy as a *middle way* solution to greenhouse gas emissions.
>
> (Nisbet, 2009: 13)

Similar trends to those identified in this quote in relation to the US can be seen in the UK and other European countries.

While the Fukushima Daiichi nuclear power plant accident, caused by the powerful earthquake and tsunami off the east coast of Japan in 2011, heralded further turbulence and upheaval in public discourse and opinion with regard to nuclear power, the negative reframing of nuclear power was nowhere near as powerful as that which had followed the Chernobyl nuclear power plant accident in 1986 (Kristiansen, 2017). Weart (2012) offers, as one of the key reasons for this, the fading (from public view and visibility) link between nuclear weapons imagery and nuclear energy following the end of the Cold War.

Do an online search using the search term 'nuclear power and climate change' (or equivalent terms or search combinations); identify from the first 2–3 pages of hits, a selection of relevant news and other Internet items (e.g. pressure group statements, blogs, etc.) that refer to or discuss the role of nuclear power in relation to public and political concern about action on climate change.

Which major 'cultural packages' (e.g. *progress, economic development/competitiveness, scientific/technical uncertainty, runaway science/Pandora's box, public accountability, middle way,* etc.) are in evidence?

Comment: note that a package such as the progress package in current discourse may well 'look' slightly different from its main articulation in public discourse of, say, the 1960s: For example, the 'progress package' traditionally cast nuclear power generation and technology as a symbol of modernity and society's progress; in its current version, it is probably more likely to appear simply as affirmation of the belief that technological innovation and prowess is the best solution to tackling and moving forward on problems such as climate change.

Which packages seem to be most prominent in the current discourse on nuclear power and climate change?

Comment: note particularly that the relative prominence of different packages does not equate with or translate in any simple way into stance, that is, the *progress package* or the *middle way* package may be prominent both because they are being used to promote nuclear power and/or because they are being used/referenced in arguments against nuclear power.

Significant changes over time in the relative prominence of different cultural narratives and interpretations of nature and the environment are also the central focus of Glenda Wall's (1999) illuminating study of a rather different genre (from science fiction film, that is), namely television documentary. Wall examines the changing discourses of science, nature and environment in the Canadian documentary series *The Nature of Things* over the extended period from 1960 to 1994. Her analysis shows that a bioeconomic outlook dominated the 1960s, that is, a perspective which saw nature as an exploitable source of resources and wealth, a domain to be studied, understood, controlled and managed by science, and a place 'where one could go to renew oneself and escape the alienating effects of city life' (p.61). This view changed gradually, moving in the 1970s towards an increasing emphasis on nature 'as vulnerable and fragile, with parts of it being under attack as a result of technological growth' (p.64), and with an increasing focus on the complexity evident in nature. At the beginning of the 1990s, 'the idea that nature will respond with a vengeance to the abuses piled upon it' (p.68) became prominent.

Wall's longitudinal analysis thus shows how discourses of nature in the television documentary changed, from the 1960s to the early 1990s, from a view of nature as a resource to be controlled and exploited, to a view of nature as fragile, but also potentially vengeful, and as deserving of respect and protection.

But while nature and the environment may have featured prominently for a long time in particular science or nature-oriented types of programming (e.g. the Canadian *The Nature of Things* analysed by Wall, or long-running documentary series such as the American *Nova* or the British *Horizon,* or indeed – as discussed in more detail below – a long tradition of wildlife films and nature programmes), analyses of mainstream television entertainment programming show a very different picture. James Shanahan and his colleagues, in a series of studies (Shanahan, 1996; Shanahan & McComas, 1997, 1999; McComas et al., 2001; see also Shanahan et al., 2015) of the portrayal of nature, the environment and environmental issues in television entertainment programmes broadcast on the main American television networks in the 1990s, thus found not only that popular television entertainment programming gave little attention overall to the environment as an issue or a problem, but also that such attention declined considerably and rapidly after the peak of environmental public concern in the very early part of the 1990s.

Nature and the environment were thus rarely – in television entertainment programming – in the foreground as a key theme or narrative component, that is, rarely foregrounded as something to be discussed or problematised. Moreover, Shanahan and McComas found that, while environmental issues and those speaking up on environmental issues may have been portrayed with some conferral of legitimacy in the early 1990s, by the mid-1990s environmental claims and claims-makers had either been absorbed within the general consumerist ideology of television or had been marginalised, 'acquired a bit of the air of the lunatic' as they put it (1999: 103).

Viewers of entertainment television programmes are, in the words of Shanahan and McComas (1999: 102–103) encouraged to see the environment as (in no particular order):

- a beautiful alternative to city life
- a 'problem' to be solved through citizen action
- a political commitment for socially marginal types
- a source of jokes

- a source of trivia (a nostalgically fading issue that characterised a particular era)
- a test and challenge for human resourcefulness.

Mainstream television entertainment programming thus taps into and reinforces many of the same cultural resonances identified in relation to film and popular culture generally. But perhaps the key lesson to be drawn from studies of television entertainment programming is that, to the extent that the environment is seen as a 'problem' or a topic for debate and discussion at all, its 'meaning' is largely framed in terms which fit the overall objective of commercial television – to promote consumerism. Television entertainment programming thus offers a symbolic environment which denies legitimacy, or at least any sense of urgency, to the notion that there may be a 'problem' with how we interact with nature and the environment, while at the same time affording legitimacy to consumer culture that is potentially highly detrimental to nature and the environment.

Very similar arguments and evidence emerge from studies of Hollywood film. Moore (2015: 12–13) notes how popular Hollywood films, while seemingly addressing key environmental issues of degradation, waste, pollution, unsustainable practices, and so on, do so in a way that merely reinforces adherence to the underlying capitalist norms of growth, productivity and consumerism: 'Hollywood takes an issue that has the potential to provide serious critique of existing consumer culture and effectively removes the critique through commodification, turning the environment into simply another product in the concentrated media marketplace.'

Focusing particularly on Disney animated films, Prevot-Julliard et al. (2015), in a systematic analysis spanning some 70 years of Disney animation (from *Snow White* [1937] to *Tangled* [2010]), show a significant decline in the representation of outdoor natural environments – particularly a decline in what they describe as 'green' environments. Depictions of green or natural environments were also increasingly shown as human-controlled or influenced, and lacking in biodiversity. They argue that the trends identified in their study reinforce an increasing 'nature disconnection' that may adversely affect how younger generations, who increasingly grow up in urban environments and predominantly experience the natural environment vicariously, come to perceive and know the natural world.

The key lessons that can be drawn from the various studies and analyses referred to so far can be summarised as follows:

1. Media coverage of environmental issues, science and technology draws on, evokes and articulates readily available cultural and interpretive packages, which in turn set particular frames or boundaries for how issues are and can be discussed and understood.
2. Cultural and interpretive packages (also often referred to in the literature as, *inter alia*: scripts, schemas, frames, narratives or discourses) are, by their very nature, *always available*, but some resonate more readily than others with widely held beliefs, fears and concerns, and such resonances change over time.
3. The silences and omissions from popular depictions and narratives about the environment are potentially as important as the framing (causes and solutions) – and associated underlying ideologies, for example consumerism, activated through this – of the environment when it *is* represented.
4. While deep-seated cultural packages can be drawn upon or evoked across widely different types of media and popular culture output, their articulation and inflection varies considerably with the genre, narrative and other (e.g. journalistic and organisational) conventions of different types of media and media content.

The latter point also redirects our attention to differences between large national or international 'mainstream' media and regional/local or 'alternative' media – a distinction that, as we have seen in the previous chapter, has important implications for how environmental issues are reported, and for how 'blame' and 'solutions' are constructed. An important study by Widener and Gunter (2007) demonstrates this very clearly through an analysis of the coverage of the aftermath and environmental recovery after the Exxon Valdez oil spill in Prince William Sound in Alaska in 1989. Their analysis focuses on the '*Tundra Times,* an Alaska Native newspaper based in Anchorage and established in 1962' (p.769) and compares its coverage of the long recovery after the 1989 oil spill with coverage in Alaskan mainstream newspapers (*The Anchorage Daily News* and *Anchorage Times*).

Using Gamson and Modigliani's notions of interpretive packages and cultural resonances, they demonstrate a clear hierarchy of source use, that is, that the mainstream media draw on traditional authority sources, while the *Tundra Times* gives proportionally more space and access to the voices of both Native and non-Native 'ordinary' people. And Widener and Gunter go on to show how *who* gets quoted has direct implications for the relative prominence of different interpretive packages and related cultural resonances.

> In this case, scientists sought precise measures of wildlife decline, reflective of a scientific worldview that shapes the image of news as objective. In our examination of the *Anchorage Daily News*, coverage of recovery was confined principally to measurable entities such as financial compensation and wildlife. [...]
>
> In contrast, the *Tundra Times* sought to connect damage with broader themes of respect, duties, and the betrayal of trust from the oil industry and the state government to the people of Alaska. For example, the 'Voice of the Times' claimed primarily that 'objective science' had been lost following the spill, indicating that scientific research that demonstrated prolonged damage lacked objectivity. On the other hand, the *Tundra Times* never limited the debate to such narrow terms. Rather, the focus in the *Tundra Times* was on the cultural meaning of the spill through a wide range of devices, including stylistic features such as poems and fables.
>
> (Widener & Gunter, 2007: 775)

The mainstream newspapers thus give priority to a science-based, expert-driven, quantitative and 'objective' measurement-oriented discourse, while the alternative *Tundra Times* affords prominence to an interpretive package that stresses the subjective lived experiences of people affected, loss of trust and quality of life issues, as well as deploying a more general 'rhetoric of loss' (see also the discussion in Chapter 2) – a rhetoric which activates notions of a past pristine idyllic utopia and pits this against notions of a damaged, spoilt, 'broken', dysfunctional present and future. The 'alternative' interpretive package, prominent in the *Tundra Times*, was further activated and consolidated through the use of culturally resonant symbols/metaphors such as 'Mother Earth' versus 'Father Oil', 'Black Death' and 'Holocaust', as well as comparisons of the oil spill with the AIDS epidemic 'to draw attention to the massive and tragic nature of their loss' (Widener & Gunter, 2007: 778).

Widener and Gunter's confirmation of the well-known finding from a large body of media research on source access and source hierarchies, that 'alternative' voices and marginalised groups are largely excluded from large mainstream media but may find a platform in smaller regional/local or 'alternative' media, points to an important dilemma for claims-makers seeking to gain a hearing for their claims in the public sphere: 'Marginalized groups turn to alternative media because their voices are excluded from mainstream media, while in turn the very "unconventionality" of their voices prohibits a connection to the mainstream' (Widener & Gunter, 2007: 780).

This suggests that if marginalised groups wish to gain access to mainstream media, not just to alternative media, they must rhetorically *repackage* their claims in terms that resonate with and conform to the dominant culture and dominant interpretive packages. But, we need to ask, can such *repackaging* be done without loss of the fundamental and defining strands and principles of a package?

The point about access to alternative media is also relevant to considerations – see Chapter 3 – about how environmental activists and pressure groups might exploit new media technologies and forms of communication for campaigning purposes. Essentially, successful claims-making is not merely or even predominantly about securing access to just any medium or line of communication, but it is crucially about 'breaking into' dominant discourses and shifting dominant frames. This is often done more effectively by *engaging* with dominant interpretive packages (particularly through subtle acts of rhetorical redefinition or reframing, including through narrative forms such as satire or comedy), rather than by merely placing alternative frames or interpretive packages alongside dominant packages.

Box 5.1

Word-matters/words matter: changing lexis/metaphor to change public connotations

Words matter. A single, well-chosen word is often enough to evoke an entire readily available cultural script or frame (see Turney, 1998, at the beginning of this chapter). There is no such thing as neutral language, just as there is 'no such thing as unframed information' (Nisbet & Newman, 2015: 325). Communication – whether in scientific articles in peer-reviewed journals, or in news reports or press releases – is in some way or form about persuasion. Environmental issues are not – and perhaps can never be – described in a completely value-free or objective fashion. As argued in previous chapters, claims-makers (whether scientists, environmental activists, politicians or simply concerned 'lay' citizens) and media professionals are fully aware of the importance of lexical choice and are fully conscious that successful communication in the public sphere is an act that requires careful attention, not just to the general thematic content or the general legal, ethical, scientific and economic, 'rights' and 'wrongs' pertaining to particular issues, but also crucially to the connotations and cultural scripts that are or may be evoked through the choice of vocabulary, the choice of particular words for describing the issue and for making claims about it.

Public and media communication about environmental issues is not then so much about information, or even about 'facts' or 'evidence' (although of course referencing 'facts' and 'evidence' is itself a rhetorical strategy aimed at enhancing the authority or credibility of what is being communicated) as it is about competition between different claims-makers and between different claims or constructions of these issues. The language used, the choice of vocabulary, is key to what we may call the ideological management of competing discourses in the public sphere. The vocabulary generally and the metaphors more specifically chosen to describe existing and newly emerging environmental problems are thus anything but arbitrary: they are, to use Günther Kress's (1997) term, highly 'motivated'. That is, they are of course deliberately chosen with a view to making them intelligible beyond a specialist audience by connoting or evoking some phenomenon, perspective or explanatory frame that is already widely known or familiar, but more importantly with a view to framing what are often contentious and controversial issues in such a way as to promote and strengthen particular arguments, interests and discourses.

When scientists and others first started drawing attention to the thinning of the ozone layer, something that was and is wholly invisible to the naked eye, the term 'hole' in the ozone layer was much more effective in raising public alarm and concern because it resonated much more widely with public perceptions and fears, and was simply easier to comprehend, than the process of 'thinning' (which of course raises difficult-to-answer questions such as 'how much' or 'how fast', 'how widespread'?; Ungar, 1992, 2003; Mazur & Lee, 1993).

It is of course not only the choice of words that is important to successful claims-making, but also – crucially for an issue which could not be visualised in conventional terms by filming it directly – the scientific graphs and computer graphics 'designed' to visualise the problem, and the wider visual icons and symbols used similarly for the purpose of reinforcing particular interpretations (Hansen, 2018).

The construction of climate change in the public sphere has a similarly interesting linguistic history. When claims-making about the potentially catastrophic outcome of rapidly rising carbon dioxide levels in the atmosphere started in the 1980s, this phenomenon was of course widely referred to as the 'greenhouse effect' and as 'global warming'. The term 'the greenhouse effect' was particularly evocative, because of its instant appeal to something that virtually everybody could relate to as a direct sensory experience, that is, humidity and heat, and although the word-image could potentially evoke positive connotations of lush and fertile growth, it also potentially carried the negative connotations of unpleasant and stifling humidity and heat.

While the terms 'greenhouse effect' and 'global warming' may have been effective lexical choices for 'sounding the global alarm' (Mazur & Lee, 1993), it also became increasingly clear during the 1990s that these were rather misleading terms that failed to capture the diverse impacts of climate change and maybe even were counterproductive in terms of raising public awareness and invoking political action. Thus, the term global warming, for example, potentially handicapped claims-makers who argued that climate change was anthropogenic, and instead afforded room for counterclaimants who argued that climate change was merely a 'natural' part of the Earth's long history of cycles of colder and warmer periods.

'Global warming' and 'greenhouse effect', in claims-making terms, were also disadvantaged by the fact that the period preceding the rise of concern and claims-making about climate change in the late 1980s, had seen a degree of prominence given to Cold War-related scientific concerns about a 'nuclear winter', the notion that a major nuclear exchange between the then superpowers would cause enough dust and particle pollution in the atmosphere to shut out sunshine and send the earth into a prolonged period of much colder temperatures. The fact that the terms 'global warming' and more particularly 'greenhouse effect' have receded from prominence in public debate since the mid-to-late 1990s is thus an example of what discourse analysts call successful 're-lexicalisation', a deliberate change of terms for the purpose of promoting a particular view, understanding or perspective (while also, in this case, removing the basis on which much of the counterclaims were built).

But perhaps one of the most active domains of 're-lexicalisation' – indicative of the deep-seated uncertainties and public anxieties which characterise this field – is that of the new genetics, of biotechnology and the manipulation of genes/genetic material in both plants and animals. Bauer et al. (1999) thus describe the significant lexical changes, symptomatic both of public/cultural sensitivities and of claims-maker interests:

> In the early days, the term biotechnology itself was hardly used. Instead, the English-speaking world commonly referred either to 'genetic engineering' or – in more technical discourse – to 'recombinant DNA (rDNA) technology'. With time, however, what came to be perceived as the negative connotations of 'genetic engineering' led to the introduction of two new terms: first 'genetic manipulation', and then (as this term, too, came to be viewed with suspicion) 'genetic modification' (GM). Recently, in what may be a borrowing from the German-speaking world, there has been a noticeable increase in the use of the term 'gene technology' (Gentechnologie).
>
> (217)

Longitudinal studies showing changes in vocabulary and indeed changes in the meaning/connotations associated with particular key terms (Condit et al., 2002, for example, track the changing meaning of the word 'mutation' in genetics coverage) help sensitise us to the idea that media and public discourse on the environment is very much actively 'constructed' and the result of deliberate rhetorical, linguistic and framing 'work' undertaken by stakeholders involved in shaping public discourse and promoting particular arguments, definitions and interpretations.

Wildlife films and nature documentaries: reading popular constructions of nature and the environment culturally

Wildlife films and nature documentaries may at first sight appear to belong to the more pleasant, wholesome, aesthetic, innocently entertaining and educational end of popular culture, far removed from the harsh, politicised, controversy-saturated and often violent realities of, for example, everyday mainstream news output. Such a perception would not be entirely accidental, as indeed it has often been one of the major explicitly stated objectives of producers of wildlife films and nature documentaries to project exactly such an image. In his impressive historical tour of American wildlife film and programming, historian Greg Mitman (1999) thus quotes from the stated objectives of the producers of the popular wildlife series *Wild Kingdom* of the late 1950s and early 1960s, that it would carry 'no vestige of political, ideological, or governmental conflict or controversy' (Mitman, 1999: 155–156). Bousé (1998) similarly notes the 'absence of overt history or politics' as one of the defining features of the much more recent mode of nature programming known as 'blue-chip' programmes (see Box 5.2).

The particular relevance of Mitman's historical analysis of wildlife and nature programming to the arguments of this chapter concern his clear exposition of how such programmes provide a carefully 'constructed' view of nature and the environment, a view which both resonates with and in turn impacts on wider social and cultural perceptions and attitudes regarding life in general and nature/the environment in particular. Because of the longitudinal and historical scope of his analysis, Mitman is also able to indicate some of the dialectical and interactive nature of the relationship between media packages, public climates of opinion and wider cultural packages.

In a very simplified form one could sketch these dialectics along the following lines: wildlife films and nature documentaries are popular with their audiences to the extent that they succeed in 'speaking to' and – implicitly – engaging with deeper public anxieties, worries and concerns; wildlife films and nature documentaries project, channel and reconstruct wider abstract social and cultural concerns within a frame which offers particular core values and world views as 'solutions' and exemplary modes of conduct. Over longer periods of time, the frequent repetition of such core values and world views impacts on public views – and ultimately on political and policy debate in the public sphere – by providing first and foremost a discursive frame or interpretive packages which define what is and can be said about, for example, nature/environment protection and conservation versus exploitation of nature/the environment (I shall return, in Chapter 7, to a more extended discussion of the conceptualisation of media 'influence' or roles inherent in this type of model).

Mitman persuasively demonstrates how some of the most popular wildlife films and nature series of the 1950s – many of them created by Disney – essentially reinforced and reworked the romantic view of nature from the 18th and 19th centuries to engage with, accommodate and reinforce dominant American cultural values of the 1950s. Mitman (1999) particularly reads Disney's nature films of the early 1950s as speaking directly and soothingly to the painful public memories of war:

> To a wider public, Disney's nature – benevolent and pure – captured the emotional beauty of nature's grand design, eased the memories of the death and destruction of the previous decade, and affirmed the importance of America as one nation under God.
>
> (110)

> Disney not only captured the aesthetic beauty of nature, he transformed it into a commodity with a set of values pertaining to democracy and morality that appealed to the American public.
>
> (124)

Not only then did wildlife and nature films of the time allow a public, horrified by the atrocities, violence and devastation of the Second World War to immerse itself, in an almost escapist fashion, in a highly selective and aestheticised construction of nature as spectacle, but the 'nature-world' on display at the same time metaphorically emphasised and affirmed a particular ideological view of society. Disney in particular used nature as a frame into which to project core American values, while at the same time lending to those very same values the power and legitimacy of 'naturalness'. In the wake of the destruction and violence of the Second World War, and amidst growing public anxiety about Communism and the escalating Cold War, Disney, argues Mitman, 'offered a persuasive vision of the American way of life rooted in individualism, traditional family values, and religious morality' (1999: 129). The supreme ideological achievement of the selective and particular construction of nature in the wildlife and nature films of the 1950s was indeed to *naturalise* core values built around the nuclear family and religion, traditional gender roles, democracy and the social order of the time.

Mitman demonstrates how this particular visualisation and construction of nature was anything but natural, in the sense that it presented a highly selective portrait of nature and the environment, a construction with a very clear convergence of particular ideological, commercial and not least ethnic and class-specific interests. Nature, in the Disney films and other popular programmes of the 1950s and early 1960s, was a sanitised version of nature emphasising the environment and nature as beautiful, idyllic, harmonious and, above all, 'pristine' – untouched and unspoilt by humans – and carefully devoid of graphic portrayals of animal violence and of any direct focus on animal sex.

> Like pristine nature, childhood, conceived as a time of innocence, offered a place of grace from the horrific acts of destruction and degenerative influences wrought by modern civilisation. Both were hallowed spaces in American society that needed to be preserved. Much of natural history films were the respectable alternative to less wholesome films in the early motion picture industry, animal shows on 1950s television offered entertaining and educational subject matter that the whole family could enjoy. In television shows like *Zoo Parade*, the construction of sentimental nature and childhood in postwar American society were closely intertwined.
>
> (135)

> In sheltering young, be they animal or human, from nuclear annihilation, the threat of communism, and the more insidious side of commercial culture, Americans in the 1950s upheld the nuclear family as a safe haven. Animal behaviour stories, especially those that focused on themes such as courtship, nest-building, parenting, and development of the young, universalised the family as a natural unit. The ideal of universality conformed precisely to the marketing needs of national television advertisers, who sought to project an image of the white, middle-class American family in which ethnic and class differences were homogenized.
>
> (141–142)

The appreciation and popularisation of wildlife and nature engendered through the 1950s' aesthetically pleasing images of a pristine and natural order, and through the construction of nature as the natural home of core cultural values, received a further twist towards domestication of animals/wildlife and spectacle in the 1960s. Mitman notes how the rise of pet-keeping and the growth of tourism in the 1960s was exploited by wildlife and nature film producers, with further implications for the campaigning strategies of environmentalists. Thus, programmes worked to establish associations between wild animals and domestic pets, and animals were imbued with personality and charisma in anthropomorphic fashion, in ways which lent themselves not only to the use of animal characters for marketing purposes, but to the marketing of 'the value of wildlife' (Mitman, 1999: 153).

> By transforming wildlife into domestic pets, animal rights groups and environmental organisations could appeal to pet owners in America [...] for support. In the 1960s, as the environmental movement gained political ground, footage and logos of charismatic species such as harp seals, dolphins, and panda bears became common in the political campaign strategies of animal rights groups and environmental organisations such as the International Fund for Animals, Greenpeace and the World Wildlife Fund.
>
> (Mitman, 1999: 154)

By the late 1970s, the sanitised and harmonious constructions of nature in perfect balance had long since given way to an altogether more violent, sensational, dramatic and dramatised depiction of nature. The historically and culturally rooted change – and the dialectic interplay, alluded to earlier in this section, between media constructions of wildlife and nature, public tastes and demands, and social and political values – can then be broadly summarised as follows:

> The immediate postwar years saw an explosive growth in nature audiences [...]. War-weary Americans looked to nature for wholesome entertainment for their children. When the baby-boom generation came of age, these childhood experiences became the seeds of 1960s environmental activism. The environmental movement in turn sparked even greater audience demand for natural history shows. [...] By the mid to late 1970s, however, television networks filled programming slots with less expensive and equally popular game shows as costs for more sophisticated technological productions of nature programs kept pace with viewers' increasingly exacting taste for more dramatic, hyperreal scenes of wildlife.
>
> [...] Furthermore, by dramatizing sex and violence found throughout the animal kingdom, such shows as *Fangs, Predators,* and weekly specials like *Shark Week* engaged new generations of enthusiasts with graphic, close-up scenes of animals copulating or predators killing prey.
>
> (Mitman, 1999: 205)

Box 5.2

Nature as narrative – narrativising nature

In Box 5.1, we looked at the key significance and power of individual terms or words in evoking or triggering chains of connotations and associated meanings. Here, I wish to draw on Bousé (1998) to move beyond the paradigmatic or word-level of analysis to point to the equally important role of the syntagmatic or narrative level as a conveyor of ideological meanings in media texts.

Bousé argues that the defining masterstroke of Disney's contribution to wild-life films was the act of imposing a conventionalised narrative framework upon them, including the use of dramatic and comic plots often reflecting familiar mythic patterns deeply ingrained in Western cultural traditions (p.130). As Bousé points out, the demand of media organisations in the 1990s continues to be for 'well-plotted, dramatic storylines and strong character development', a success-ful formula that was well established in the Disney productions of the 1960s and has since been exported and adapted worldwide, including in British natural history films.

> This 'classic format' for wildlife films, with its 'rules' and narrative conventions, has also been flexible and adaptable enough over the years to accommodate stories about domesticated animals, from Disney's *The Incredible Journey* (1963), to its recent Japanese rip-off *The Adventures of Milo and Otis* (genially narrated by Dudley Moore). It has also survived intact without voice-over narration in films like Jean-Jacques Annaud's *The Bear* (1989), and Disney's *Homeward Bound* (1993; also a reworking of *The Incredible Journey*).
>
> (Bousé, 1998: 132)

The core narrative formula thus runs along the following lines – note the similarity with the classic folk/fairy-tale formula identified by structuralists such as Vladimir Propp (1968), Roland Barthes (1977) and others:

> The story typically opens in spring, and centers on a protagonist (some-time more than one) who is either orphaned, abandoned, or separated from family or community. It then faces a perilous journey or struggle to survive so that the lost primal unity of family and/or community can be regained. Present in this 'family romance' are such mythic narrative ele-ments as departure, separation, quest, initiation, and triumphant return or re-union. Though the perilous journey motif has given way in recent years to the perilous ordeal, such as weathering the dry season, or surviving attacks by predators (or humans), the relation of these events to the theme of youthful initiation remains central.
>
> (Bousé, 1998: 132)

As a result, largely of economic and media organisational pressures to maximise audiences, the tendency today, argues Bousé, is in the direction of what has become known as 'blue-chip' films. In terms of 'constructions of nature', perhaps the key point that stands out is the way in which several of the core features combine to project and reinforce an image of nature as a pristine and unspoilt 'Eden on today's Earth', devoid of the politics, controversies, problems and stresses of modern civilisation, yet serving, through the narrative and anthropomorphic depiction of animals, to 'naturalise' and reinforce particular cultural and social values and arrangements.

Bousé (1998: 134) offers this helpful characterisation of the key features of 'blue-chip' nature programming:

1. *The depiction of mega-fauna* – lions, leopards, tigers, bears, sharks, and other large predators – although elephants, whales and few other non-predators are also included;
2. *Visual splendor* – magnificent scenery, beautiful sunsets, and stunning panoramas as a background to the animals, all of which suggest a still unspoiled, primeval wilderness;
3. *Dramatic narrative* – this can entail the classic, animal protagonist-centered narrative, or some version of the 'family romance', or even a narrative centering on the film-maker's encounter with the animals, but in any case usually includes some dramatic chases and escapes;
4. The *absence of history and politics* – no overt (...) propaganda on behalf of conservation issues and their causes;
5. *The absence of people* – although the filmmaker can occasionally appear as a character to provide the point-of-view, more than one or two people can spell the introduction of scientific and technical conservation efforts that spoil the 'natural' picture;
6. *The absence of science* – while perhaps the weakest and most often broken of these 'rules', the discourse of science can entail its own narrative of research (see Silverstone, 1984), with all its attendant technical jargon and seemingly arcane methodologies, which, like history and politics, spoil the picture of nature in all its 'natural' splendor.

Thanks to the excellent – often longitudinal and historically focused – analyses of wildlife films and nature documentaries by the likes of Mitman (1999), Bousé (1998, 2000), Davies (2000a, b), Aldridge and Dingwall (2003), Cottle (2004), Dingwall and Aldridge (2006) and others, we now have a relatively comprehensive understanding of the narrative formats, key values and associated constructions of nature and the environment in this popular genre, particularly for the post-war period and up until the late 1990s. As I have argued above, one of the most interesting aspects of analysing wildlife films and nature documentaries concerns their ideological content. That is, the way in which they both express and ideologically reinforce particular and selected social values and world views, while also in the longer term contributing to changes in social and public perceptions and attitudes to nature. What is fascinating is the intricate web of interaction between material and social realities, their representation and articulation in media and popular culture, and changes in public and personal views about the environment, including about how we relate to, treat, use and protect nature and the environment in ways that are both 'sustainable' and (ethically) responsible.

EXERCISE 5.2 Constructing nature and naturalising social/cultural values?

Analyses (including those by Mitman and Bousé referred to above) of wildlife films and nature documentaries have emphasised the way in which the narrative and format conventions of these types of programmes require and work towards a view of nature as separate and distinct from mankind or civilisation. The illusion created for us is nature programmes as a 'window' on nature, a passive observer of nature and animals going about their business. Major nature documentaries of the last two decades or so would appear, however, to have deliberately broken with this traditional mould by devoting, for example, the last ten minutes of an hour-long episode to a 'behind-the-scenes' sequence focusing on the patience, stamina and technological ingenuity required by camera crews and on the trial-and-error process of capturing nature in what Bousé (above) refers to as its 'natural splendour'.

Consider one of the most prestigious, most expensive and most widely viewed nature documentary series of recent times, *Planet Earth* (2006) and its sequel ten years later, *Planet Earth II* (2016), produced by the BBC and presented and narrated by Sir David Attenborough.

Watch one or more episodes from each of the two series of *Planet Earth* and explore how far the various characteristics delineated in this chapter, for example with reference to Mitman and Bousé's analyses, apply.

In relation to the questions listed below, consider particularly what changes are evident between the first series, *Planet Earth,* and the more recent *Planet Earth II*, for example in narrative format and in terms of depictions and messages about nature, wildlife and environment:

- Does the sheer 'visual splendour' suggest 'unspoiled, primeval wilderness'?
- Are people and other signs of civilisation's intrusion absent? Is the impression conveyed that we as viewers are simply spectators, voyeurs 'looking in' on nature?
- Does the 'behind-the-scenes' addition at the end of each episode effectively alert us, the viewers, to the constructedness of what we see? Does it matter if it does or doesn't?
  Comment: it would be fair to assume that one reason why a series such as *Planet Earth* has been so successful is precisely because it excels on all the normal criteria (technical, visual, aesthetic, narrative, etc.) and viewer expectations of what a nature documentary should do and should look like. The implicit criticism in much of the research on nature documentaries seems to be that these conventions carry with them and perpetuate an idealised – maybe romanticised – and environmentally perhaps 'unrealistic' view of our natural environment.
- What kinds of storytelling or narrative devices are used?
  Comment: note particularly the importance of visuals, sound/music and, of course crucially, the voice-over narration which tells us what the visuals 'mean' and provides narrative continuity.
- Does the narrative structure/formula conform – in whole or in part – to the 'core narrative formula' described above with reference to Bousé?
  Comment: does the narrative formula follow the seasonal (spring, summer, etc.) and/or life cycle of a central protagonist (an individual animal, a nuclear family or larger unit of a particular species, etc.), construed along such key narrative elements as 'departure, separation, quest, initiation and triumphant return or reunion' (Bousé, 1998: 132)?
- What values or ideologies are communicated through the narrative structure and the visual/verbal characterisation of wildlife?
  Comment: are animals imbued – for example, through the voice-over narration – with human characteristics and values? If so, which values and can these values be seen to relate to a particular social class, religion, age- or ethnic group, or culture?

- Depending on the answer to the previous question, is it possible to argue that the programme effectively 'naturalises' or 'universalises' values and views which are perhaps not universal but rather specific to particular social, ethnic, cultural or philosophical groups?

While several studies have examined and commented on the very considerable technological, narrative and other format changes which have impacted on the genre of wildlife and nature programming since the 1990s, we know rather less than is the case for the period from the 1940s to the late 1980s about how or whether these changes have significantly altered the ideological messages of this important and popular genre, or indeed in a wider sense, engaged with and/or influenced public and popular views of the environment and nature.

Advances in film, television and visual technology seemingly went hand-in-hand with changes in public tastes and public demands of the genre of wildlife and nature programmes to move away from the sanitised narratives of the 1950s and early 1960s towards, as Mitman and Bousé and others have noted, a more graphic, dramatic and raw mode of portrayal. But while this change could be seen at face value as a move away from the highly constructed and structured narrative form in the Disney-mode of the 1950s and 1960s towards a more realistic and realist – perhaps more documentary-like and factual – mode, the real change is more likely away from the realist illusion of the earlier decades to a postmodernist preoccupation with spectacle and multiple representational modes, with little or no concern for 'facts' or 'reality'.

Bagust (2008), drawing extensively on Scott (2003), thus notes:

> For Scott (2003), the 'computer generated extravaganzas' of the *Walking with …* documentary 'franchise' represent the nature film (sub-?)genre's desperate attempt to maintain a foothold in an increasingly global, fragmented and profit-driven television market. To do this, she suggests, producers are increasingly looking away from traditional 'blue-chip' wildlife film conventions (with their pretence that they access 'unmediated reality') towards a product that jumps generic boundaries and 'ups the representational ante' through the use of high-tech visual effects to create a level of spectacle which 'produce[s] a visceral response in the viewer by way of the sheer audacity of the image itself' (Scott 2003, 30).
>
> (219)

If the emphasis of the postmodernist wildlife and nature programmes is on the visually spectacular and impressive, rather than on conventional storytelling or the communicative construction of nature in terms of recognisable and culturally widely shared value sets or packages, how does this impact on public perceptions, attitudes and actions with regard to nature and the environment?

While emphasis can be placed on the notion that nature and wildlife documentaries, driven by visual and competitive imperatives in the increasingly diverse media environment, are becoming less truthful, accurate or realistic – for example, through the use of extensive reconstructions of, for example, animal action or hunt scenes (San Deogracias & Mateos-Pérez, 2013) – this possibly misses the key point. Thus, what matters most in all mediated communication and in human communication generally is not so much benchmarking against some ill-defined standard of accuracy (whose 'accuracy'?) or an equally elusive standard of truthful or 'realistic' representation/portrayal. Rather, it is perhaps the notions of authenticity (Kirby, 2014) and of narrative coherence that matter. Nature documentaries and wildlife films then are judged, not primarily on the basis of how accurate or realistic they are, but on the basis of whether they 'work' in terms of narrative coherence, good storytelling and are visually attractive.

Conclusion

Moving beyond the specific focus on *news* reporting of the environment, this chapter has examined the notions of scripts, cultural packages, interpretive packages and cultural resonance, and explored their significance for understanding the nature and potential 'power' of popular media representations of nature and the environment. We have seen how media representations of environmental issues, science and technology draw on, evoke and articulate readily available cultural and interpretive packages, which in turn set particular frames or boundaries for how issues are and can be discussed and understood in the public sphere.

Cultural and interpretive packages (also often referred to in the literature as, *inter alia*: scripts, schemas, frames, narratives or discourses) are, by their very nature, *always available*, but some resonate more readily than others with widely held beliefs, fears and concerns, and such resonances change over time in response to changes in socio-historical conditions as well as in economic pressures, organisational arrangements and technology (including media technology).

While deep-seated cultural packages can be drawn upon or evoked across widely different types of media and popular culture output, their articulation and inflection varies considerably with the genre, narrative and other (e.g. journalistic and organisational) conventions of different types of media and media content. As we have seen in this chapter, much is known about narrative and stylistic formats of wildlife films and nature programming from the post-war period till the 1990s, *and* about the wider cultural values and views of nature and the environment evoked and expressed in these. While a growing body of research has charted many of the significant changes of the more recent period in narrative and stylistic formats, in 'character' development and in thematic content, there is still much work to be done on the question of how these changes have impacted on the media and public ideological constructions of nature and the environment.

The following chapter continues the examination of how constructions of nature and the environment change over time, but pursues the investigation by focusing particularly on another important media genre: advertising.

Further reading

Moore, E. E. (2017). *Landscape and the Environment in Hollywood Film: The Green Machine*. Palgrave Macmillan.

Ellen Moore provides one of the most comprehensive and insightful studies to date of the portrayal of the environment and environmental issues in popular Hollywood film. Demonstrating the significant variations across different film genres, she shows how surface imagery stressing the seriousness of environmental problems often draws from and reinforces underlying messages/ideologies, such as consumerism, that run counter to environmental sustainability.

Nisbet, M. C., & Newman, T. P. (2015). Framing, the media, and environmental communication. In A. Hansen & R. Cox (Eds.), *The Routledge Handbook of Environment and Communication* (pp. 325–338). London and New York: Routledge.

Nisbet and Newman provide an exemplary update for the present century, focused on the relevance to environmental communication research of framing and of Gamson and Modigliani's classic and influential 'cultural packages' framework.

Bousé, D. (2000). *Wildlife Films*. Philadelphia: University of Pennsylvania Press.

Like Mitman's (1999) *Reel Nature*, Bousé provides an impressive and comprehensive analysis of the long history of wildlife and natural history film-making with a keen focus on production dimensions as well as on the conventions and ideological messages of the films themselves.

6 Selling 'nature/the natural'

Advertising, nature, national identity, nostalgia and the environmental image

THIS CHAPTER:

- Discusses how concepts of 'nature' are historically specific and constructed.
- Explores the ideological uses of nature in advertising and other media output.
- Shows that while environmental themes and 'green' advertising come and go, appeals to nature and the natural have featured prominently in advertising since the early days of advertising itself.
- Examines the key 'uses' of nature in advertising and the extent to which they reflect changing historical constructions of nature, social change and changing views of society.
- Explores media uses of nature in relation to the concepts of nostalgia and national/cultural identity.
- Examines whether increasing globalisation in advertising is at odds with culturally specific concepts/images of nature and environment.

Introduction

Nature imagery and ideas regarding 'the natural' form, as we have seen in previous chapters, are an important part of media and popular culture generally. Such images are also prevalent in the particular media genre of advertising. While appeals to environmental protection and 'green' or 'sustainable' consumption wax and wane in public communication, nature imagery continues to figure prominently in commercial advertising and other media discourse. Advertising is perhaps particularly interesting in this regard because it effortlessly and seamlessly draws on culturally deep-seated, ontological and taken-for-granted meanings of nature and the natural, and reworks these in ways which promote consumption, particular world views and particular identities.

Starting from the recognition that nature is socially constructed, and that its social construction changes over time, this chapter explores the key ways in which advertising and other media use and articulate ideas of nature. There is much literature to suggest that constructions of nature are nationally and culturally specific. There is also evidence from the communications research literature to suggest that media constructions of nature – including in advertising – often significantly draw on and in turn reinforce notions of national and cultural identity. These arguments and suggestions are examined in light of the trends of globalisation and homogenisation in advertising.

The chapter examines the nature discourses deployed in advertising content, the extent to which these resonate with environmental concerns, and the extent to which they articulate and reinforce culturally specific identities. It is argued that advertising, in its uses of nature, makes an important contribution to ongoing public definitions of the environment, consumption and cultural identity. I examine how discourses of nature change over time and explore how advertising articulates and reworks deep-seated cultural categories and understandings of nature, the natural and the environment, and in doing so, communicates important boundaries and public definitions of appropriate consumption and uses of the natural environment.

Constructed nature and ideology

A discussion about nature and identity in advertising perhaps needs to start by reiterating the simple recognition of the complex and historically changing meanings associated with 'nature'. Not only does nature figure prominently in our cultural vocabulary and in popular culture, but it means different things, at different times to different people. Cultural critic Raymond Williams called nature 'perhaps the most complex word in the language' (Williams, 1983: 219), and he noted how dominant cultural views of nature have changed over time, from, for example, the Enlightenment view of nature as something to be studied, understood and controlled, to Romanticism with its emphasis on nature as pure, spiritual, sublime, authentic and pristine. Particularly relevant to studying the uses of nature in advertising and other media is Williams's observation that 'one of the most powerful uses of nature, since the late

18th century, has been in this selective sense of goodness and innocence. Nature has meant "the countryside", the "unspoiled places", plants and creatures other than man' (1983: 223).

The power and semantic complexity of nature derives not only from the fact that dominant interpretations of nature change historically, but perhaps even more so from the contemporaneous coexistence – within the range of meanings associated with the word – of a rich cultural reservoir of binary opposites: 'nature was at once innocent, unprovided, sure, unsure, fruitful, destructive, a pure force and tainted and cursed' (Williams, 1983: 222). Referencing or using nature thus offers potentially rich interpretative flexibility (to the extent that we as members of a culture have access to the repertoire of meanings culturally and conventionally associated with nature), while at the same time appearing to render things ontological, permanent and beyond questioning. References to nature or what is regarded as 'natural' are key rhetorical devices of ideology in the sense that they serve to hide what are essentially partisan arguments and interests and to invest them with moral or universal authority and legitimacy.

The ideological power of nature referencing derives, as Armitage (2003) – drawing on Williamson (1978) – has argued, from the archetypal binary opposites of culture versus nature:

> What is 'natural' has a dual, even contradictory, role in advertising. Nature is given meaning through culture, but its significance lies in the fact that nature is understood to be outside or even the very opposite of culture. The cultural messages attached to nature thereby appear to have an authority independent of and superior to any particular culture.
> (Armitage, 2003: 75)

Contrary to its surface appearance of being beyond culture, beyond that which has been culturally created and constructed for particular ideological purposes, the appearance of referencing something stable, permanent, authentic, inviolable and God-given, nature can be, and has been, used for lending authority to particular ideas, interests and political objectives. Evernden (1989: 164), with reference to Marshall Sahlins (1977), refers to 'the general and possibly essential propensity of human societies to invent the nature they desire or need, and then to use it to justify the social pattern they have developed'. Nature, Evernden continues, 'is used habitually to justify and legitimate the actions we wish to regard as normal, and the behaviour we choose to impose on each other.'

Thompson (1990: 66), in his comprehensive discussion of ideology, culture and media, refers to this as the 'strategy of naturalisation' and others, like Fairclough (1989) and Stuart Hall (1982), have similarly noted the ideological character and centrality of a discourse of naturalisation in media and political rhetoric. Stuart Hall offers a particularly succinct observation of the way in which television, both as 'content' and as (visual) 'medium', hides its own highly selective representation of 'reality' through a process of naturalisation:

> Much of television's power to signify lay in its visual and documentary character – its inscription of itself as merely a 'window on the world', showing things as they really are. Its propositions and explanations were underpinned by this grounding of its discourse in 'the real' – in the evidence of one's eyes. Its discourse therefore appeared peculiarly a naturalistic discourse of fact, statement and description. But […] it would be more appropriate to define the typical discourse of this medium not as naturalistic but as *naturalized*: not grounded in nature but producing nature as a sort of guarantee of its truth.
>
> (Hall, 1982: 75)

If nature referencing and naturalisation are key rhetorical components of the way in which ideology is communicated, then the semiotic linking of a (romanticised) view of nature *with* a rural (idyllic) past *with* national identity has undoubtedly been one of the most potent ideological uses in the modern age, utilised in the early parts of the 20th century for naked political propaganda and mobilisation for war, and in the second half of the 20th century for commercial purposes:

> More insistent in recent times has been the commercial recourse to the 'rural imaginary', in order to reconcile us to the activities of multinational companies and to solicit our custom for their commodities. Today, it is the marketing rather than the political propagandist potential of nature that is more exploited, and the clichés of nationalist rhetoric have become the eco-lect of the advertising copywriter.
>
> (Soper, 1995: 194)

The use of nature and referencing of the natural in advertising may be partly to do with ideological power (selling products or corporate images by invoking the qualities of goodness, purity, authenticity, genuineness, non-negotiability (Cronon, 1995)) and partly to do with the format constraints of the advertising genre. Thus, the format and time constraints of advertising, and indeed of social media such as Twitter, call for the use of easily identifiable and recognisable shorthand symbols, or what Gamson and Modigliani (1989) aptly refer to as 'condensing symbols', and nature and the natural are perhaps amongst the most universally recognisable of such symbols. This is not to say that appeals to nature and what is regarded as natural are insignificant in other genres; quite to the contrary, they play a powerful role in much media and public discourse, including in public debate and controversy around biotechnology and genetics (Hansen, 2006), abortion, sexuality, same-sex marriage, race and gender roles, and so on. But it is to suggest that their uses in advertising may be more frequent and perhaps also more heavily stereotypical in the sense of relying on a narrower and more limited range of essential 'nature/natural' icons.

What particularly distinguishes the use of nature in advertising is the seamless way in which, in its predominant use, it blends in 'naturally' (for want of a better expression) and almost unnoticeably. It is thus its relative inconspicuousness, combined with the deep-seated – culturally and historically specific – values/views which its uses represent, that gives nature its ideological power. The ability of advertising to forge signification links which convey such key nature-related values as freshness and health onto cigarettes and smoking (Williamson, 1978) is perhaps one of the clearest examples of the semiotic flexibility and power of uses of nature in advertising. Nor has the potential power of nature referencing been lost on corporate image advertisers keen to enhance connotations with environmentally responsible, ethical, sustainable behaviour.

Environment and nature in advertising and other media

Numerous studies have documented the considerable increase in news media reporting (and public concern) about environmental issues which happened in the latter half of the 1980s and very early 1990s (Hansen, 2015a; Boykoff et al., 2015). A number of studies have, likewise, begun to reveal some of the characteristics and trends in representations of the environment, nature or environmental issues in advertising. Comparing a sample of American television advertisements from 1979 with a sample a decade on, 1989, Peterson (1991) thus found that explicit environmental or ecological messages were used in less than a tenth of advertisements, although slightly more frequently in 1989 than in 1979. Several studies from the mid-1990s (e.g. Banerjee et al., 1995; Beder,

1997; Buckley & Vogt, 1996; Iyer & Banerjee, 1993; Kilbourne, 1995) indicate that the increased news and public interest in the environment, seen in the late 1980s, was also reflected in advertising, where so-called 'green advertising' or 'green marketing' became prominent. Much of this advertising latched on to general public concerns by labelling products as 'green' and 'eco-friendly' or by emphasising measures taken to reduce the potential impact that advertised products might have on the environment. The trend also extended to corporate image advertising stressing the environmental credentials of large companies, and to advertising more directly using the tradition of persuasive information/education campaigns in the form of government and local authority advertising campaigns designed to promote recycling initiatives or environmentally responsible behaviour (Rutherford, 2000).

Much research in this century on images/messages regarding the environment has focused on 'green' advertising/marketing and on 'greenwashing', interpreted as the use of deceptive claims or disinformation 'regarding the environmental practices of a company or the environmental benefits of a product or service' (Baum, 2012: 423). In a study of print advertisements in leading news, business, environmental and science magazines in the USA and UK in 2008, Lauren Baum (2012) found three quarters of 'environmental' advertisements to contain one or more aspects of greenwashing, and she argues that without stricter regulations, a trend of increasing use of green or environmental appeals – and of greenwashing – in product and corporate image advertising is set to continue unchecked.

The overall trend indicated here is confirmed by a comprehensive study by Lee Ahern and colleagues (Ahern et al., 2012; Bortree et al., 2013) of environmental messages in advertisements in *The National Geographic* magazine over the three decades from 1979 to 2008. Ahern and colleagues show a significant increase in corporate environmental responsibility communication over those thirty years, as well as considerable changes in the framing of environmental messages. They particularly note an increasing emphasis on promoting the environmentally positive actions ('doing more') of corporations rather than an emphasis on conservation ('taking less from the Earth').

In an exemplary earlier study, Howlett and Raglon (1992) chart the changing uses of appeals to nature and 'the natural' in advertising during the 20th century. They show that while product association with nature and the natural changes little, 'environmental' corporate image advertisements gain momentum principally since the early 1970s, as companies start to 'portray themselves as nature's caretakers; environmentally friendly, responsible, and caring', resulting, as they continue, 'in the use of more scenes from nature (...) and an almost total elimination of the factory and machinery visuals which were standard fare in the corporate image ads of the 1950s' (Howlett & Raglon, 1992: 55).

These and other studies (e.g. Plec & Pettenger, 2012; Atkinson & Kim, 2015; Segev et al., 2016) then confirm the continued and increasing reference to environment, nature and environmental issues in advertising, the prominence of corporate voices in such advertising, and the increasing use of frames which show products and corporations as environmentally friendly and responsible, as nature's caretakers, and as innovative leaders in environmental protection (Rutherford, 2000; Schlichting, 2013).

EXERCISE 6.1 Packaging corporate business as Nature's Caretaker

Take a look at the websites of selected major oil companies (e.g. BP, Shell or Exxon Mobil) and consider:

1. To what extent are the images used to represent/symbolise the environment 'iconic' and global images of pristine nature, lush nature, untouched nature? Is nature principally shown 'on its own' and separate from humans or human artefacts?
2. Is there discursive and ideological reconciliation between, on the one hand, the environment/nature as a resource to be mastered and controlled, and, on the other, the environment/nature as a fragile and precious system to be protected? Consider particularly the significance of language use, for example the deliberate choice of words such as 'protect', 'support', 'responsible', 'responding', 'regenerate', 'restore', 'stewardship', 'respect', 'sustainable', 'helping', 'local', 'community', 'families', 'people', and so on.
3. Which ideological frames or packages – for example progress, economic development, morality/ethics, public accountability, and so on (see also the discussion of core frames/packages in Chapter 5) – are most prominent?

Box 6.1

Regulating green advertising

Building on research in the 1990s into the referencing of green, sustainable, environmental and ecological credentials in product advertising, a number of studies in this century have examined the green or environmental claims that advertisers make for their products. Segev et al. (2016) thus conclude, on the basis of a comprehensive content analysis of magazine advertising, that advertisers are increasingly responding to growing public concern for the environment, and that 'the majority of environmental claims were deemed acceptable, implying a trend toward more trustworthy and reliable green advertising'. By contrast, Atkinson and Kim (2015) – in a study that confirms the prominence of green-ad frames across a range of product advertising – conclude that so-called green claims are generally 'ambiguous and unsubstantiated'. Atkinson (2017: 10) goes further to argue that 'the preponderance of evidence seems to suggest misleading and deceptive claims continue to plague the advertising genre'. What these and other studies point to is the increasing concern about labelling standards and practices, which often make it difficult for consumers to assess the quality or veracity of green advertising claims, and the lack of or limitations of regulation in this field. Baum's (2012) comparative study of green advertising claims in the UK and the USA, finding deceptive or misleading claims to be considerably more prominent in the USA than in the UK, points to the importance of considering the regulatory environment.

In the UK, the Government's Department for Environment, Food and Rural Affairs (DEFRA, 2016) offers guidance on making 'an environmental claim for your product, service or organisation' (www.gov.uk/government/publications/make-a-green-claim/make-an-environmental-claim-for-your-product-service-or-organisation). The document also lists the relevant legal framework/legislation as follows:

> 7. Relevant legislation
> Before making green claims, you should check if the law requires you to state or publish environmental information about your product. You should check the following legislation:
>
> - EU Unfair Commercial Practices Directive (UCPD)
> - The Consumer Protection from Unfair Trading Regulations 2008 (CPRs)
> - The Business Protection from Misleading Marketing Regulations 2008 (BPRs)
> - The Sale and Supply of Goods to Consumers Regulations 2002).

The document importantly notes that 'DEFRA has no enforcement role on environmental claims except for the EU Ecolabel' and then goes on to list 'Other bodies [which] have a role in enforcing or regulating environmental claims', including the Advertising Standards Authority (ASA) and the Committee of Advertising Practice (CAP). The Advertising Standards Authority is thus in charge of dealing 'with complaints about advertisements, promotions and broadcast adverts. ASA makes sure CAP and BCAP advertising standards codes https://www.asa.org.uk/codes-and-rulings/advertising-codes.html are applied'.

While the prominence of explicit appeals to 'environmentally responsible' behaviour in advertising goes up and down in ways seemingly not dissimilar to the ups and downs seen in news media attention to environmental issues (see Chapter 2), there is considerable evidence (e.g. Howlett & Raglon, 1992; Goldman & Papson, 1996; Rutherford, 2000; Corbett, 2002; Hansen, 2002; Ahern et al., 2012; Atkinson & Kim, 2015; Segev et al., 2016) that references to nature and the natural have been prominent throughout the 20th century and continue to be so in the 21st century. Goldman and Papson (1996) persuasively argue that:

> From its inception, modern advertising used nature as a referent system from which to derive signifiers for constructing signs. Nature's landscapes were used to signify experiences or qualities that urban-industrial life failed to provide. [...] Advertising suggested that civilisation's deficiencies could be ameliorated by consuming commodities that contained the essence of nature.
>
> (191)

Drawing on a number of historical analyses (including Marchand, 1985), they argue that nature was prominently used in advertising in the 1920s and 1930s in a nostalgic way that was itself a response to the economic and the social–psychological crisis of the 1920s. Tracing the general trends in uses of nature in advertising up through the 20th century, Goldman and Papson note that '[t]he nostalgia for nature evident in the advertising of the 1920s and 1930s gave way to the fetish of gadgetry' (p.191) for the middle decades of the 20th century, and not until the 1970s did nature once again take a central position in advertising. While nature is thus seen to have been prominent in advertising throughout the 20th century, Goldman and Papson also point out that the genre known as 'green' advertising did not emerge until the 1980s.

Taking his point of departure in the now famous 'Crying Indian' advertisements of the early 1970s, historian Kevin Armitage (2003) shows how (American) popular culture has long used the stereotype of the noble savage to epitomise and articulate a nostalgic view of unspoilt nature and of an idealised past of harmony between man and nature. Armitage persuasively argues that 'The fascination with nature and the primitive that marked turn-of-the-twentieth-century American culture was rooted in a larger ambivalence about modern life' (p.73) – not unlike the disillusion with modernisation identified by Raymond Williams (1973) in the British context – but crucially, as Armitage argues, the fascination with nature and the primitive 'did not involve a rejection of civilisation, but rather an accommodation to modern life that was simultaneously nostalgic and progressive, secular yet spiritually vital' (pp.73–74). Armitage shows that the idealised referencing of nature – through or with representations of the American Indian – in advertising was well under way towards the end of the 19th century.

In an exemplary analysis of advertisements for pesticides in agricultural magazines, spanning the half-century from the 1940s to the 1990s, Kroma and Flora (2003) demonstrate the changing prominence of three different discourses: in the 1940s–1960s, a 'science' discourse articulating the post-war faith in progress through science; in the 1970s–1980s, a 'control' (of nature/the environment) discourse drawing extensively from military/combat control metaphors; through, in the 1990s, a 'nature-attuned' discourse reflecting environmental sensibilities – concerns about sustainability, protection of and harmony with nature – emerging in the latter half of the 20th century. They conclude that 'changing images reflect how the agricultural industry strategically repositions itself to sustain market and corporate profit by co-opting dominant cultural themes at specific historical moments in media advertising' (Kroma & Flora, 2003: 21).

Wernick's (1997) comparison of 1950s and 1990s advertising adds further confirmation of the changes in the ways in which nature imagery has been deployed in advertising. Where the 1950s advertisements celebrate 'the fruits of industrial civilisation' (p.209), gadgetry, technology, science and progress, the 1990s adverts appeal to nostalgic ideas of nature and the past, a nature and a past that exist only in myth: 'The Good identified as the essence of its product is located in the past, not in the future. It (…) is something to be recovered rather than attained' (p.210).

Drawing on the arguments presented by these authors, it seems then that a key difference in the uses of nature between advertising of the 1940s–1970s and advertising of the late 20th century is one of perspective: the adverts of the middle part of the 20th century in short *look forward*, with optimism even, to the progress and prosperity of the techno-scientific urban society, while the perspective of the late 20th century is one of *looking back* – to recover a lost idyll, harmony, authenticity and identity of a (mythical) past. Wernick refers to the 'progress myth' of the 1950s advertising; others, notably Gamson and Modigliani (1989) in their study of the framing of nuclear power since the mid-20th century in popular culture and in public opinion, refer to this as the 'progress package' – a common and prominent frame in media and popular culture accounts involving the relationship between technology and nature, and valorising technological, economic and scientific progress above concerns for the environment or nature. Wilson (1992) likewise identifies this period as one in which the relationship with nature was one of domination and greed, where the urge to 'acquire and consume' (p.14) far outpaced any hint of concern about the environment, limited resources or the protection of nature.

Rutherford (2000) introduces his own label for advertising celebrating the progress myth: 'Technopia'. In his historical sweep of what he broadly terms 'advocacy advertising', he implies a similar trend to that identified by Wernick, Goldman and Papson and others. He contrasts the 'Technopia' type of advertising – advertising which principally promotes a belief in the scientific and technological control and domination of nature as synonymous with progress and development – with what he terms 'Green Nightmare' advertising. 'Green Nightmare' advertising emphasises and calls public attention to the 'Dystopia' – the destruction of nature, the environment and our entire habitat – resulting from the unchecked and wasteful production and consumption practices characteristic of late modernity.

In very general terms, Rutherford's analysis maps onto the time line indicated above, namely with 'Technopia' advertising most prominent in the 1960s–1980s, and 'Green Nightmare' advertising prominent from the 1970s onward. If Rutherford's categories seem to overlap considerably, it is perhaps confirmation, not only of a diversification of discourses on nature, but of a public sphere marked increasingly by discursive competition over the framing and meaning of nature generally, and more specifically of the framing of science, technology and progress in relation to public conceptions of nature.

In summary, then, it would seem from the work of the authors discussed above that nature imagery has been a feature of advertising since at least the early part of the 20th century. It is also clear that the particular deployment and constructions of nature in advertising and other media have, broadly speaking, oscillated between, at the one extreme, a progress-package-driven view of nature as a resource to be dominated, exploited and consumed, and, at the other extreme, a romanticised – and often retrospective – view of nature as the (divine) source and embodiment of authenticity, sanity and goodness, to be revered and protected (or to use a more current invocation: 'not to be tampered with').

Nature, identity and nostalgia

Nature imagery in advertising of the late 20th century is, as we have seen above, often deployed in relation to a retrospective look, a yearning for the 'idyllic past'. Nature imagery in this context is used to construct a mythical image of the past (including childhood) as a time of endless summers, sunny and orderly green landscapes, and, perhaps most importantly of all, as a time and place of community, belonging and well-defined identity. Several researchers have referred to this view as one of 'nostalgia':

> Nostalgia became, in short, the means for holding onto and reaffirming identities which had been badly bruised by the turmoil of the times. In the 'collective search for identity' which is the hallmark of this postindustrial epoch – a search that in its constant soul-churning extrudes a thousand different fashions, ecstasies, salvations, and utopias – nostalgia looks backward rather than forward, for the familiar rather than the novel, for certainty rather than discovery.
> (Davis, 1979: 107–108)

The nostalgic view of the past, as enacted through the use of nature imagery in advertising, is not merely a longing for a mythic past, but it is very much also a romanticised view of the past. In its use of nature imagery, it draws particularly on the romantic view of the countryside, the view constructed not least by the poets (e.g. William Wordsworth, Samuel Taylor Coleridge) and painters (e.g. John Constable, J.M.W. Turner) of the Romantic period. As Williams (1973) has pointed out, the growing cultural importance of a romanticised view of the countryside perhaps not surprisingly coincided – at least in Britain – with a period of immense social upheaval, urbanisation, migration to the cities and the rapid decline of a rural/agrarian economy.

Against the tremendous social, economic and political upheaval characterising much of the 20th century, not least the first half, it seems perhaps hardly surprising that advertising should respond with romanticised images of a more natural, rural, countryside past, where identities seemed more firmly fixed, if only through 'knowing one's place' in the highly hierarchical structure of rural society. What is particularly ideological about this reconstructed past is the way in which the deeply hierarchical structures are either glossed over *or* romanticised and portrayed as indeed natural, desirable and harmonious.

The romanticised construction of nature and the uses of idyllic nature in advertising are then not just a matter of advertising responding to a public sense of alienation or a public search for identity. They are an ideological reconstruction in the sense that they naturalise, and sometimes even celebrate, a deeply stratified society. There are, in other words, important social class, race and gender dimensions to these uses of nature.

Phillips et al. (2001) in their analysis of the construction of rural/countryside/nature imagery in British rural television drama thus show that the dominant construction of a rural idyll goes further to 'also enact particular social identities, including, but not exclusively, those of class' and that the class identity enacted is predominantly a middle-class identity. Others (Scutt & Bonnet, 1996; Thomas, 1995) have similarly commented on the social class, race and gender dimensions of television and print media constructions of countryside and nature, and their associations with 'Englishness':

> In order for rural areas to continue to represent Englishness, it is essential that they are ideologically and physically separated from those groups who do not conform to stereotypical images of the English. Social and racial exclusion can, therefore, be seen as intrinsic to the maintenance of the countryside as a cultural reservoir.
>
> (Scutt & Bonnet, 1996: 8)

In a similar vein, but with a particular focus on racial exclusion in American magazine advertising from 1984–2000, Martin (2004) points to the racial dimension of nature imagery: 'Advertisements taking place in the Great Outdoors or featuring models participating in wilderness leisure activities rarely include Black models, while advertisements featuring White models regularly make use of Great Outdoors settings and activities' (p.513). He notes that the dominant view of nature/wilderness in advertising is a white Eurocentric view which finds little resonance amongst Black and Native American audiences. Similar conclusions regarding the under-representation of non-whites in the promotion of nature and outdoors recreation are also reached by Kloek and her colleagues (2017: 1033) in a comprehensive visual analysis of pictures in magazines and on websites of four large Dutch nature conservation organisations. They found that only 3.8% of the depicted people were non-white and concluded that this type of imagery (re)produces an image of outdoor recreation and nature conservation as activities performed almost exclusively by whites.

Nature and national identity

A considerable body of literature has pointed to the links between particular constructions of nature and national identity. Macnaghten and Urry (1998), drawing on a broad range of work from geographers, sociologists and historians, note how every nation celebrates its particular nature. 'National natures' may not seem particularly 'constructed' where these bear a seemingly obvious relation to the particularly striking features of those natures (the Alps of Switzerland, the fjords of Norway, the forests and lakes of Finland, etc.). However, on closer historical scrutiny, it becomes clear that 'national natures' are indeed very much 'constructed'. This is made particularly clear in historian Simon Schama's insightful analysis of *Landscape and Memory* (Schama, 1995), in which he demonstrates the particular historicity, political role and construction of nature in the culture and politics of a range of nations (with examples ranging as widely as the 'forest' in German culture and history to 'wilderness' and national parks in the United States). Geographers, historians, sociologists and media researchers in Britain have commented on the close links forged, from the 1800s onwards, between national identity and a romantic view of nature. More specifically, this is a view which constructs the 'green and pleasant land' of William Blake's famous hymn, *Jerusalem*, as both the true home and the essence of Englishness:

> Between 1880 and 1920 the conviction that English culture was to be found in the past was stabilised. Images of rolling hills, winding lanes, country cottages and the concept of an organic community of days of yore were offered as the source of English culture and the physical setting of ideologies of Englishness.
>
> (Scutt & Bonnet, 1996: 5)

Since the late 1800s then, the dominant image of Englishness in literature, art and popular culture generally has become one of equating

Englishness with the countryside, the countryside as the true home of the English (seen as white and middle-class) and the essence of Englishness. However, as Thomas (2002) amongst others have noted, the 'association of national identity with a country's rural roots is not confined to Britain, and may be connected to the cultural homogenisation which is one of the outcomes of globalisation' (p.34).

Following a similar line of argument, Creighton (1997), looking at domestic tourism and popular culture in Japan, demonstrates a renewed search for authentic Japanese identity as manifested in the increasing popularity of 'traditional' rural Japan. She describes the 'retro boom' – a looking back to the past and a search for authentic Japanese identity – experienced in Japan since the 1970s as a reaction to 'the perceived threat of cultural loss to which the processes of modernisation and Westernisation have subjected modern Japan' (p.242). As in British advertising and popular culture, the 'place' of authentic national culture is seen as the countryside or traditional village: 'This retro boom has romanticized Japan's agrarian heritage, allowing the domestic tourist industry to capitalize on travel promotions featuring rural hamlets that were formerly considered unsophisticated and boring' (p.241).

For the alienated urban masses, the search for identity is perhaps answered through the travel – or pilgrimage, even – 'back' to the true time and place of Japanese culture and identity, the romanticised rustic countryside setting. But, as Creighton demonstrates, this journey is increasingly commodified in popular culture, department store displays and consumer goods, so that the busy urban dweller need never leave the city in order to consume – or buy into – the retro boom construction of Japanese cultural identity.

As in the West, the achievement of advertising deploying this kind of nature imagery is to channel the yearning for authenticity or identity or the pure goodness of nature into consumption: purchasing the advertised product becomes a means of 'buying into' the identity or the authenticity ostensibly anchored in the idyllic rural past.

While, as indicated by Macnaghten and Urry (1998) and others, there are different 'national natures', it is perhaps testimony yet again to the semantic flexibility hinted at by Williams (1983), when he suggested that 'Nature is perhaps the most complex word in the language', that some have implied a degree of global universality in 'nature imagery' and cultural constructions of nature. Howlett and Raglon (1992) thus note that:

> Elements of nature have long provided humans with the symbols and metaphors that have helped order and explain the world, and these symbols and metaphors are extraordinarily resilient, long-lived, and in some cases at least, seem even to transcend specific cultural and linguistic borders.
>
> (60)

In their view then, the attractiveness of nature imagery and symbolism to advertisers stems from the simple recognition that '[N]atural symbols and metaphors are among any culture's most easily understood ones' and they 'tend to be long-lived and their meanings widely accessible' (p.61). The similarities noted above between British and Japanese linking of national identity with a (romanticised) rural, idyllic, countryside past likewise suggest an element of culture-transcending universality. In a comparative study of cultural values in American and Japanese advertising, Barbara Mueller (1987) found that on the two nature-related dimensions investigated ('oneness with nature appeals' and 'manipulation of nature appeals') there was remarkably little difference between the advertising of the two countries:

> Themes emphasizing the goodness of nature are found in both Japanese and American advertisements. A subtle difference in the application of this appeal exists: U.S. advertisements in this category focus on natural as opposed to man-made goods, while the Japanese advertisements emphasize the individual's relationship with nature.
>
> (56)

In an interesting follow-up study some 20 years later, Okazaki and Mueller (2008), comparing findings for 1978 advertising with 2005 advertising in Japan and the USA, found the 'oneness with nature appeal', identified as more characteristic of Eastern culture, continued to be prominent in Japanese advertising, but had dropped significantly in American advertising. By contrast, 'manipulation of nature appeals' had increased from a mere 0.7% (1978) to 2.6% (2005) in Japanese advertising, but were rarely deployed in US advertising of either period. The overall conclusion from the follow-up study was that advertising appeals in the two countries had become increasingly similar.

In her study of cultural values in Chinese and American television advertising, Carolyn Lin (2001) notes that previous studies have shown that 'advertisements in China are more likely than Western advertisements to use appeals of traditional values such as status and oneness with nature, whereas U.S. advertisements reflect such values as individualism and manipulation or control over nature' (pp.86–87). Her own study likewise confirms that Chinese advertisements are more likely to use oneness with nature appeals than US advertising.

Cho et al. (1999) identified three categories of appeals involving nature in American and Korean advertising: Manipulation of Nature, Oneness with Nature and Subjugation to Nature. They found that only Oneness with Nature featured prominently in the advertising of both countries, while Manipulation of Nature and Subjugation to Nature were comparatively rarely used. While the differences between the two countries were not statistically significant, Cho et al. found 'non-significant directional support for the contention that oneness-with-nature is found more often in Korean commercials and that manipulation-of-nature is found more often in U.S. commercials' (p.68). Another study of Korean advertising, in a comparison with Hong Kong advertising (Moon & Chan, 2005), found a greater – but still relatively small – use of appeals to the natural in Korean advertising than in Hong Kong advertising, but, as in the study by Cho et al., the difference was not statistically significant.

There is, however, considerable evidence from elsewhere (e.g. Kellert, 1995) to suggest that the uses and interpretation of nature do indeed vary across different cultures, that nature is indeed not only culturally constructed but also culturally specific in its construction/interpretation. Any apparent similarity, across (Occidental and Oriental) cultures, in advertising and other popular culture constructions/uses of nature are thus more likely to be symptomatic of the increasing globalisation, Westernisation and homogenisation, characteristic of modern advertising trends, than of some universality of nature as a sign and metaphor.

A particularly interesting – but somewhat different – inflection of nature and national identity in advertising is the use of national stereotypes. Advertising, as Armitage (2003) has shown, has long articulated and exploited the popular culture stereotype of the American Indian as the idealised 'child of nature', epitomising a nostalgic anti-modern sentiment and a yearning for a lost harmony between man and nature. But this type of inflection also extends beyond peoples (the American Indian) to national stereotypes. A particularly potent example is the referencing of the Irish and Irishness in the global marketing of Irish Spring soap by the American Colgate-Palmolive company. Elbro (1983) and Negra (2001) both offer insightful analyses of the ways in which the Irish Spring advertisements draw on and reinforce the (stereotypical) linking between nature (pure, cleansing and untainted by modernity), the (idyllic) past, nation and national (Irish) identity. As in much other popular culture construction of national identity (see Creighton on Japan, referred to above), nostalgia plays a key role, in that the linking or association also implies that Ireland is a place where the (natural) qualities of the past can still be found, visited and consumed, or alternatively, bought into through consumption of the advertised product.

The Irish Spring soap advertisements have used, for example, idyllic images of the Irish countryside (winding lanes, hedges, fresh and green) and the Irish (jolly courting couples in rural attire) to associate the qualities of freshness, authenticity, genuineness, romance, and so on, with the advertised product. The advertisements trade on and reinforce a nostalgic stereotypical image of Ireland as a 'non-industrialized paradise populated by simple country folk' (Negra, 2001: 86) and of Irishness as synonymous with honest, authentic, natural, uncomplicated, pure and romantic qualities. Condensing-symbols of nation and national identity linked to landscape and the natural environment are of course widely exploited in tourism advertising (see, for example, Urry, 1995 and 2001; Negra, 2001; Nelson, 2005; Clancy, 2011; Uggla and Olausson, 2013), and very likely increasingly so, as the popularity of 'nature-tourism' and 'eco-tourism' continue to rise.

Although there is a large and growing body of comparative research on the cultural values reflected in advertising across different cultures, only a small number of these have touched specifically on nature imagery or uses/constructions of nature in such advertising, and this is clearly a field ripe for further investigation. Research needs to compare the uses/constructions of nature and environment in television and other media advertising of several different countries. It further needs to examine how far, in a time of increasing globalisation (not least in marketing and advertising), constructions of nature are either increasingly universal or homogeneous (or Western perhaps), or alternatively, how far advertisers in their use of nature imagery seek to reflect and appeal to regionally, nationally or culturally specific understandings of nature.

Conclusion

Raymond Williams argued that 'nature is perhaps the most complex word in the language' – the present review of how nature has been used and appropriated in media discourse essentially confirms this. It testifies to the signifying flexibility of nature, and to the historically changing underlying views of nature. It is only by examining how discourses of nature change over time that we can begin to understand how they are used ideologically for promoting everything from national identity, nationalism, consumerism and corporate identity to framing and circumscribing what kinds of questions can and should be asked about the environment, environmental issues and (sustainable) development.

Invoking nature/the natural in advertising and other public discourse is a key rhetoricial device of ideology in the sense that referencing something as 'of nature' or as 'natural' serves to hide what are essentially partisan arguments and interests and to invest them with moral or universal authority and legitimacy. Nature, as Evernden (1989: 164) so succinctly puts it, 'is used habitually to justify and legitimate the actions we wish to regard as normal, and the behaviour we choose to impose on each other'.

While explicit environmental messages and 'green' advertising seem to come and go, with some periods of prominence in advertising since the early 1970s, it is clear from the literature examined here that nature and appeals to the natural have been a significant part of advertising since at least as far back as the early 1900s.

The particular deployment and constructions of nature in advertising and other media have, broadly speaking, oscillated between, at the one extreme, a progress-package-driven view of nature as a resource to be dominated, exploited and consumed, and, at the other extreme, a romanticised – and often retrospective – view of nature as the (divine) source and embodiment of authenticity, sanity and goodness, to be revered and protected, 'not to be tampered with'.

On the continuum between these extremes lie, as we have seen, a wide range of 'constructions' of nature, including: nature as resource, good, authentic, idyllic, healthy, spiritual, enchanting, the 'home' of identity, fragile, a threat, a 'proving ground' for both human and product qualities, vengeful, and so on. Several authors have indicated that the dominant construction of nature in advertising and popular culture of the late 20th century is one which draws heavily on nature imagery of the Romantic period. It is also one which invokes a nostalgic view of the past, with implications for the public construction of social class, gender, race and, not least, national identity.

There are contrasting – and possibly contradictory – indications from a number of studies regarding the extent to which advertising and other media constructions of nature are culturally or nationally specific. Cross-cultural comparisons of advertising show relatively small differences in the views/constructions of nature in Occidental and Oriental advertising, and indicate that traditional – and well-documented – differences in views of nature may be subsumed under the homogenising influence of globalisation. But the evidence is very tentative and further research should focus specifically on the construction of nature to examine the extent to which the discourses of nature in advertising are culturally and nationally specific.

Further reading

Williamson, J. (1978; reissue edition: 2010). *Decoding Advertisements: Ideology and Meaning in Advertising*. London: Marion Boyars.

Judith Williamson's insightful critical introduction to the analysis/decoding of how meaning and ideology are constructed in advertising continues to be one of the best and is as applicable today as it was when first published. See particularly Chapter Four: 'Cooking' Nature, and Chapter Five: Back to Nature.

Atkinson, L. (2017). Portrayal and impacts of climate change in advertising and consumer campaigns. *Oxford Research Encyclopedia of Climate Science*. Retrieved 23 Aug. 2017, from http://climatescience.oxfordre.com/view/10.1093/acrefore/9780190228620.001.0001/acrefore-9780190228620-e-376

Lucy Atkinson offers a comprehensive overview of research on the content and influences – on environmental attitudes as well as consumer behaviour – of green advertising.

Schmuck, D., Matthes, J., Naderer, B., & Beaufort, M. (2018). The effects of environmental brand attributes and nature imagery in green advertising. *Environmental Communication, 12*(3), 414–429. doi: 10.1080/17524032.2017.1308401

Schmuck and her colleagues offer an overview of experimental research on the use of nature imagery in advertising. Their experiment shows, amongst other things, that the use of nature imagery positively influences brand attitudes and purchase intention.

Stöckl, H., & Molnar, S. (2018). Eco-advertising: the linguistics and semiotics of green(-washed) persuasion. In A. F. Fill & H. Penz (Eds.), *The Routledge Handbook of Ecolinguistics* (pp. 261–276). London: Routledge.

Drawing from semiotics and the emerging field of eco-linguistics, the authors analyse example cases from three subgenres of green advertising: green commercial advertising, green-washed commercial advertising, and green non-profit/social advertising.

Media, publics, politics and environmental issues

THIS CHAPTER:

- Examines selected measures of public concern for or interest in environmental issues, noting the long-term parallels between public opinion and media coverage regarding the environment.
- Discusses some of the major approaches which have been used for researching media influence on public and political opinion and political decision-making.
- Reviews the evidence on media influence produced by major approaches such as agenda-setting research, framing analysis and cultivation analysis.
- Explores the referencing and use of 'media coverage' for rhetorical and political purposes in public environmental controversy, arguing that this can be seen as an important type of 'effect' or influence, albeit one much less commonly thought of in general discourse on media influence.
- Concludes, via a brief discussion of alternative approaches to studying media roles in relation to public understanding(s) of environmental issues, by emphasising the 'circulation of claims' perspective over a linear perspective on media roles and by emphasising the complex and multiple ways in which media coverage interacts with other 'forums of meaning-creation' (Gamson, 1988) in society.

Ultimately, the assumption, whether explicit or implicit, behind most research into media representations of environmental issues is that these play a role in shaping and influencing public understanding/opinion and political decision-making in society. Perhaps regrettably, concerns about the 'effects' or role of the media are often construed in relatively simplistic terms which assume a simple relationship between media coverage and general public belief, attitude and behaviour, while ignoring the diverse chains of influence which characterise the media's role and function in society. This problem is by no means unique to 'environmental issues' and the media, but indeed has a far longer pedigree in discussions about the media and politics, crime, violence, terrorism, international conflict, race and ethnicity, and so on.

While the tradition of studying the social role and effects of the media has long since moved away from simplistic notions of 'direct effects' of the media, public and political discourse on the media and the environment frequently appear to still operate with tacit assumptions about 'vulnerable' publics easily manipulated or swayed by 'powerful' media. The rhetoric of powerful media and vulnerable publics is indeed itself often deployed by sources and stakeholders seeking to promote their particular concerns and definitions in the public framing of political arguments about environmental issues. Although undoubtedly instances and circumstances exist where something approximating a powerful and immediate media impact on public beliefs can be found, the aim of this chapter is not to dwell on anecdotal examples, but rather to examine some of the wider research evidence on media roles in relation to the dynamics of public and political discussion and action on environmental issues.

This chapter then discusses some of the major frameworks which have been used for examining media influence on public understanding, public opinion and political decision-making. It explores the evidence, from a variety of research approaches and frameworks, for how mediated public communication about the environment influences political processes or is interpreted and engaged with by different publics.

Public opinion and the environment

Public opinion, attitudes, understanding and behaviour in relation to environmental issues have been the subject of research, as indeed has media coverage, since the emergence of the 'environment' as an issue for public and political concern in the 1960s. As noted in earlier chapters, this is of course not to say that environmental issues and problems did not exist before the 1960s, but merely to stress that the more holistic way of thinking about – and labelling – seemingly diverse issues/problems such as air, soil and water pollution, waste disposal, overpopulation, excessive exploitation of natural resources, and so on, as interconnected and part of ecology only began to crystallise in the 1960s.

General surveys of public opinion and attitudes regarding the environment have relatively consistently confirmed: 1) that, while public concern for the environment ebbs and flows in cycles over time, 'the environment' has, since the late 1960s, become firmly established as a core issue of public interest on the public and political agenda; 2) that mediated public communication is a major – in some cases *the* major – source of public information and knowledge about the science and scientific evidence regarding environmental issues (Hargreaves et al., 2004; Whitmarsh, 2015), particularly where these are new and rapidly developing issues; 3) that the extent to which publics are influenced by, draw on and/or use media information/images is influenced by socio-demographic and political characteristics (Roser-Renouf et al., 2015), local knowledge, first-hand experience and availability of other – more direct – sources of information.

Studies of public opinion and attitudes on environmental issues suggest that public awareness and concern about the environment has waxed and waned in cycles not dissimilar to the up-and-down cycles of media coverage of environmental issues, discussed in Chapter 3. In very general terms these cycles can be described as follows: public concern about the environment developed during the 1960s and reached an initial peak around 1970, then fell back during the 1970s; a second cycle started with a gradual increase in public concern during the 1980s, reaching a peak around the early 1990s, and then waning again from around 1992 onwards, while the first decades of the present century have witnessed yet another resurgence, centred in particular around an increased sense of urgency about climate change, but also in relation to more specific (albeit often related to climate change) environmental issues, such as pesticide use, organic food production, air pollution, waste and recycling practices, plastics pollution, and so on. Again, however, it is important to note that while the overall long-term trend of enhanced public awareness and concern about the environment appears to point solidly in the upward direction, the up-and-down cycles have continued during the first two decades of the 21st century. Thus, one global survey published in 2013 reported environmental concerns to be 'at record lows' (Globescan, 2013), while an American survey in early 2017 reported an increase in public concern for the protection of the environment: 'A majority of Americans (55%) now cite protecting the environment as a top priority, up from 47% a year ago' (Pew Research Center, 2017). The 2017 Eurobarometer survey of citizens of the 28 member states of the European Union similarly noted that the percentage of citizens saying that 'protection of the environment is important to them personally' had increased since the previous survey in 2014. Although there are some superficially tantalising parallels in trends since the 1960s, between the up-and-down cycles of media coverage of environmental issues and the ups and downs of public concern, as measured through opinion polls, it would be a mistake to attempt to draw any simple conclusions about cause and effect regarding the relationship of media coverage and public concern.

In an early review of evidence on public opinion and attitudes on environmental matters, Lowe and Rüdig (1986), for example, point out that longitudinal and more comprehensive surveys have shown public concern about the environment to be relatively resilient, stable and widespread. Like many others before them (e.g. Funkhouser, 1973), they suggest that the fluctuations registered by more superficial opinion polls may merely reflect 'the immediately prevailing preoccupations of the mass media' (Lowe & Rüdig, 1986: 514), rather than the public's actual concern about environmental issues.

While longitudinal studies of media coverage of a range of environmental matters have demonstrated considerable fluctuations – characteristic, as argued in earlier chapters, of the news value orientation and other factors impinging on the media communication of social issues – studies of public opinion have generally pointed to a greater degree of stability, and indeed to slow but steady increases, in public concern about the environment (Dunlap, 1991, 2006; Brulle et al., 2012; Whitmarsh, 2015).

The regularly conducted Eurobarometer surveys indicate, as suggested above, increasing concern about environmental issues amongst citizens of the European Union countries, while also showing – unsurprisingly – some considerable country-by-country as well as wider regional variations. The 2017 Eurobarometer shows (see Box 7.1) climate change and air pollution to be the two top environmental issues of concern:

> Europeans are most likely to say that climate change is one of the most important environmental issues (51%), followed by air pollution (46%) and the growing amount of waste (40%).
> More than a third consider the pollution of rivers, lakes and ground water an important issue (36%), while around a third choose the following issues: agricultural pollution and soil degradation (34%), the decline or extinction of species and habitats, and of natural ecosystems (33%) and marine pollution (33%).
>
> (Eurobarometer, 2017: 5)

Box 7.1

Attitudes of European citizens towards the environment (Eurobarometer, 2017: 5)

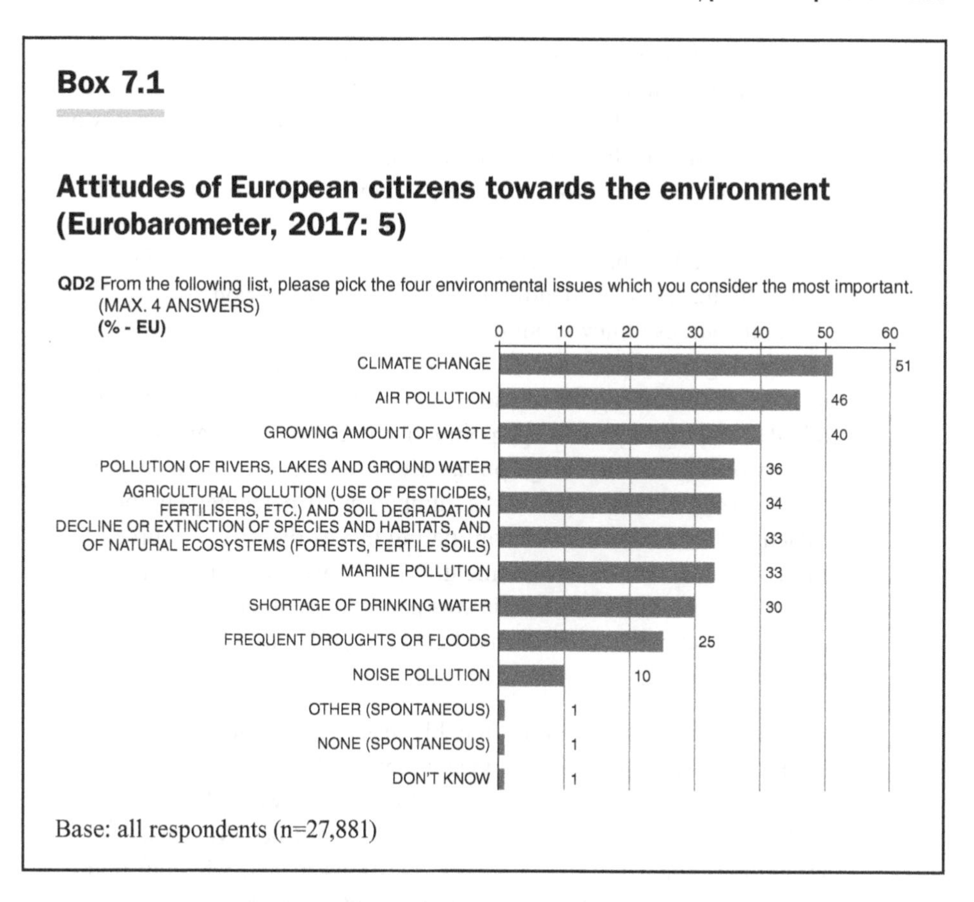

Base: all respondents (n=27,881)

The traditional way of beginning to account for the role of communications media in influencing public opinion – whether regarding environmental issues or indeed any other issues – has long since been that of asking people what their main sources of information are. While this can only ever be a starting point for beginning to understand the complexity of media influence, it nevertheless provides a useful indication of the relative importance of different types of media in our public communications environment. This type of evidence becomes even more compelling, when looked at longitudinally, providing a sharp reminder of both the size and speed with which the communications media are changing, and with them our ways and means of keeping informed about the environment and other social issues. Unsurprisingly, the evidence from regularly conducted surveys, such as the Eurobarometer surveys, is that traditional print and broadcast news media are steadily declining in importance, while the Internet and social media are becoming increasingly prominent sources. The difficulties of accounting (including in the wording of survey questions) for the increasing convergence of traditional and new media notwithstanding, what *is* perhaps surprising is the continued prominence of television news, which remains the top source of information about the environment. Also surprising and noteworthy is the continued resilience of radio news. As shown in the Eurobarometer graph in Box 7.2, newspapers, television and radio news are all steadily declining over the period 2004 to 2017 as main sources of news, but radio much less so (a drop of 6% in those mentioning radio as a main source of environmental news) than television news (remaining the top source of information mentioned at 58%, albeit a drop of 14 percentage points from 72% in 2004) and particularly newspapers (a 25 percentage points drop from 51% to 26%). By contrast, 'the internet and social networks' jumped by 31 percentage points from 11% in 2004 to 42% in 2017 nominating this as their main source of information about the environment. It is also worth noting that the rise of the Internet and social media was particularly steep in the earlier part of that period, while the increase between 2014 and 2017 was a mere one percentage point.

Although these trends unquestionably testify to a rapidly changing media and communications environment, the figures also undoubtedly need to be taken with a pinch of salt: the lines between different types and sources of information are not always clear (including to us as individuals when we answer survey questions such as what our main source of environmental information is) and indeed becoming less so as different media converge and/or increasingly feed off each other.

The gradually, but steadily declining prominence of 'films and documentaries on television' as a main source of information is perhaps surprising in light of the considerable increase in the number and diversity of 'environment-focused' documentaries and films (including what has been characterised as a new genre of climate fiction or 'cli-fi' (Svoboda, 2016)) seen since the release in 2006 of *An Inconvenient Truth*. Despite the seeming decline in prominence, it is also, however, worth noting the continued overall dominance of the television medium, especially when observing that television occupies both the top and the third most prominent spot as key source of information about the environment.

On a wider global scale, drawing on data from 38 countries around the world, the Pew Global Attitudes Survey 2017 asked respondents

Box 7.2

Main sources of information about the environment (Eurobarometer, 2017: 7)

QD3 From the following list, which are your three main sources of information about the environment?
(MAX. 3 ANSWERS)
(% - EU)

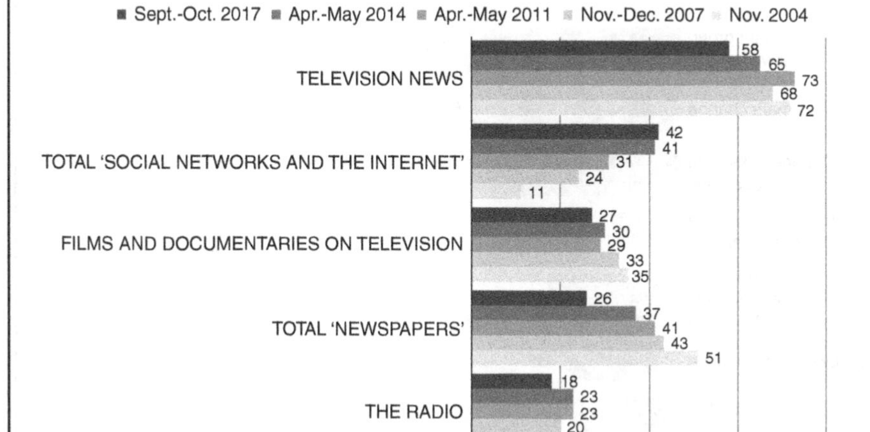

Top 5 answers
Base: all respondents (n=27,881)

what they considered to be major threats to national security. The findings show that respondents overall consider climate change as a top threat, second only to ISIS, to their country, and indeed ahead of other threats such as cyberattacks, the global economy and immigration/refugees.

While opinion poll surveys at both national and global level thus seem to confirm that public concern about the environment is both well-embedded and indeed shows signs of having increased considerably during the first decade of the new millennium, it is also clear from international surveys such as the Eurobarometer surveys and the Pew Global Attitudes surveys that huge variations exist across countries and regions of the world. Moreover, while the longer-term trends do indeed confirm that public concern about the environment is here to stay and

Box 7.3

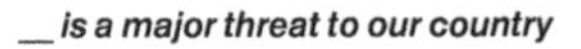

Pew Global Attitudes Survey 2017 of perceived major threats to national security (Pew, 2017: 2)

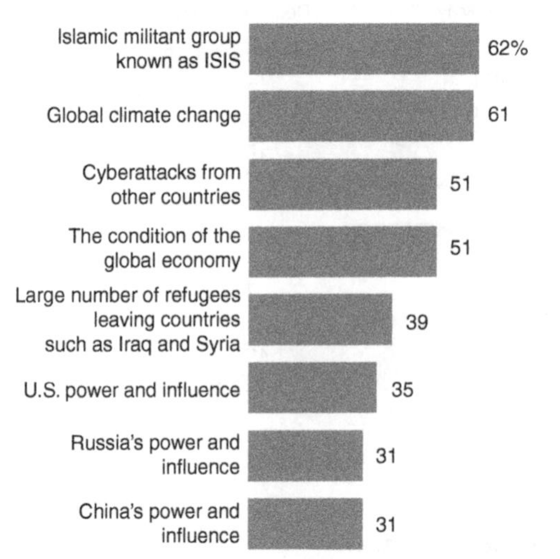

Note: Figures represent global medians across 38 countries. ISIS not asked in Turkey, U.S. power and influence not asked in U.S, and Russia's power and influence not asked in Russia.
Source: Spring 2017 Global Attitudes Survey. Q17a-h.

PEW RESEARCH CENTER

remains amongst the major issues on public agendas, we should also be aware of the continued shorter-term ups and downs of the major issues competing for public attention in the public arenas. As we have seen with regard to media agendas, issues compete with each other for prominence on the public opinion agenda, and all such arenas are furthermore characterised by their limited carrying capacities – when one issue moves up on the agenda, others have to move down or give way altogether. It is one thing to try and determine the interaction of different issues on the public opinion agenda; it is a different, and somewhat more complex, problem to try and map the interaction of different public agendas (e.g. 'public opinion', 'media coverage', 'government and political agendas' etc.). In the remainder of this chapter, I examine some of the theories and approaches which have been used for studying these relationships.

Major communications approaches to media and public opinion

While a multitude of approaches have been used, the main approaches in media and communication research which have contributed to examining the relationship between media coverage of the environment and public/political understanding of environmental issues are agenda-setting, framing, cultivation analysis and, to a lesser degree, the 'quantity of coverage' hypothesis.

Agenda-setting research

The starting point for agenda-setting research is commonly seen to be Cohen's (1963: 13) formulation that 'The press may not be successful much of the time in telling people what to think, but it is stunningly successful in telling its readers what to think *about*.' The first empirical test of the formulation, however, was not published till nearly a decade later in McCombs and Shaw's (1972) study of agenda-setting during the American presidential campaign of 1968. Interestingly, the agenda-setting paradigm thus emerged in mass communications research at around the same time as the environment and environmental issues began to emerge as issues for public and political concern, and it is perhaps not surprising, although probably quite coincidental, that agenda-setting has been a relatively prominent approach to studying the role of the media in relation to environmental issues since the late 1960s/early 1970s.

Thus, Funkhouser (1973) in his study of the 'issues of the 1960s' found a tantalisingly close match between many of the issues/news stories, including 'Environment', dominating American news magazines of the 1960s and the issues, nominated in Gallup surveys of the American public as 'the most important problem facing America', whilst also concluding that 'the news media did not give a very accurate picture of what was going on in the nation during the 1960s' (p.73).

> The data cited here suggest that the amount of media attention given to an issue strongly influences its visibility to the public. However, the amount of media attention does not seem to relate as closely to public attitudes concerning the issues and related policies. The Gallup item, 'What is the most important problem facing America?' may in effect be an indirect content analysis of the news media, showing us the surface of public opinion but not its depth.
>
> (74)

Similar trends have been confirmed in a string of agenda-setting studies since (Atwater et al., 1985; Ader, 1995; Mikami et al., 1995; Sampei and Aoyagi-Usui, 2009). These studies have shown the ability of the media to raise general public and political awareness about environmental issues, although identifying the specific ways in which media agendas and public agendas co-vary over time has proven a more challenging task. Fan et al. (1994) showed how German television news in 1986 influenced the public agenda on energy supply and other issues. Their analysis showed that television coverage of the Chernobyl accident and other issues influenced and contributed to the public agenda and helped increase public concern about energy supply.

In an earlier publication, Brosius and Kepplinger (1990), in a complex agenda-setting study in Germany, similarly found evidence of relatively strong agenda-setting effects on the issues of energy supply and environmental protection. They also, however, noted some of the complexities in interpreting the findings arising from their research: they noted, for example, the very different agenda-setting timescales applying to the development of general perspectives such as environmental protection, compared with awareness of specific events (e.g. the Chernobyl accident) within a wider issue.

Christine Ader's study, published in 1995, provides particularly strong evidence for the agenda-setting role of the media on the issue of environmental pollution. The study is noteworthy not only for the extended period of time analysed, 1970 to 1990, but also for its careful comparison of media and public agendas with 'real-world' conditions or indicators of environmental pollution. Where much agenda-setting research has thus been relatively narrowly focused on the relationship between two agendas, the media agenda (as determined through systematic content analysis) and the public agenda (as determined through surveys of public opinion), Ader convincingly argues for the need to 'control' for the possibility that any observed co-variation between the media and public agendas may be caused by a third exogenous factor.

Like researchers both before (e.g. Zucker, 1978) and since (see Soroka, 2002, discussed below), Ader points out that there are particularly good reasons for expecting the media to have an agenda-setting effect in relation to environmental issues: environmental issues are often – although certainly not always – what agenda-setting researchers call 'unobtrusive' issues in the sense that they are often not easily observed or experienced first-hand. It is this relative absence of direct observation, experience or indeed of more immediate sources of information, which makes it possible for the media to 'step in' as the main source of information (and potentially 'influence') for the public.

On the face of things, it looks as if climate change is very much not an unobtrusive issue, in the sense of agenda-setting researchers, that is, climate change is all around us for everybody to observe and experience first-hand. However, it is clearly important to remember that whatever 'symptoms' of climate change we see around us, they are of course only just that because we have been told (often through the media) that this is what they are, manifestations of climate change. It is not so long ago that a whole host of natural phenomena such as flooding, hurricanes, droughts, hot summers, and so on, would *not* automatically have triggered references to global warming or climate change in the way that has now become more or less customary.

Ader's study, comparing the press and public agendas over the period 1970 to 1990, while controlling for the impact of real-world conditions, found strong support for the agenda-setting power of the media. Thus, the study confirmed that the level of public concern about environmental pollution was influenced by the amount of media attention devoted to this issue, and further that:

> When the effects of reality were controlled, the correlation between the media agenda and the public agenda was strengthened. As predicted, this study found that real-world conditions do not influence the media or public agendas directly. The public needs the media to tell them how important an issue the environment is. Individuals do not learn this from real-world cues. Also, the media are not effective at determining the importance of this issue from real-world cues.
>
> (Ader, 1995: 310)

Where early agenda-setting studies of media coverage and environmental issues were principally focused on the relationship between amount of media coverage and issue importance as designated by public opinion surveys, more recent agenda-setting research has focused on the complex interactions among several public 'agendas', including media agendas, political/policy agendas, 'real-world cues' and public opinion.

In a longitudinal study of the agenda-setting process in relation to global warming, Trumbo (1995) found that the influence of media coverage on politicians (members of the US Congress) was considerably more pronounced than media influence on the public opinion agenda. In a particularly sophisticated design, Soroka (2002) similarly examined interactions of the agendas of Canadian newspapers, public opinion polls, formal political forums and legislative initiatives from 1985–1995. He found that the media were a powerful agenda-setter on environmental issues, but less so on other issues where the public could draw on personal experience or had access to more immediate and more trusted sources of information. Similar conclusions emerged from a longitudinal analysis in Belgium by Walgrave et al. (2008), who also found media agenda-setting on environmental issues to be stronger than on other issues.

Other longitudinal analyses have likewise found evidence of the influence of media coverage on the political and legislative forums. Thus, Jenner (2012) found that news photographs affect policy-maker attention but seem to have a more ambivalent impact on public attention to environmental issues. Also focusing on images and 'visual agenda-setting', Miller and LaPoe (2016) combined image content analysis of mediated images of the BP Deepwater Horizon oil spill disaster of 2010 with a survey of public recall of images relating to the disaster. Their study confirms the basic proposition of agenda-setting, that the most prominent media images are also those most remembered/recalled by the public. But the study further shows that visual agenda-setting is strengthened by emotion. Thus, emotions evoked by oil-soaked animals elevated that set of images to the most memorable, even if these images were not the most prominent in media content. The study confirms the power of mediated images – particularly where people's principal information source is mediated information, as was the case in relation to the Gulf of Mexico oil spill. Visuals, then, add a powerful additional dimension to media agenda-setting. As Miller and LaPoe (2016) eloquently put it:

> Visuals add to the interpretation of an event and often show a deeper meaning that results in action by the audience. That action can include mobilization, activism, and protest, which often lead to changes in procedure or policies—all because someone captured the event on a cell phone and a television station decided to run it.
>
> (61)

As in agenda-setting research generally, determining the particular direction of influence between media coverage and public opinion with regard to environmental issues has generally proved challenging. While there is thus much evidence from longitudinal studies that the media can play a potentially powerful role in influencing the public opinion agenda, there is also, however, some evidence on certain environmental issues of the reverse type of agenda-setting where public issue concerns appear to drive issue coverage in the news (Uscinski, 2009). This apparent contradiction, itself in part a product of examining agenda interaction over extended periods of time, merely confirms what communication researchers have argued for some time, that media roles are best understood – not from a simple linear perspective – but from a much more dynamic and interactive perspective that recognises the dynamic and fluid nature of interactions between key 'meaning creating forums' (Gamson & Modigliani, 1989) in society.

In one of the most complex and sophisticated tests of agenda-setting on environmental issues to date, Liu et al. (2011) examine media and congressional attention to climate change over the extended period of 1969 to 2005. Their study casts valuable light on how problem indicators (e.g. quantitative statistical indicators), high-profile international events and science feedback regarding climate change influence media and political attention. Their findings confirm that the three sets of 'attention-grabbing' factors do 'indeed generally promote issue salience', albeit to different degrees and with different timescales for media and political agendas. Indicative of the complexity of agenda-setting, they conclude that although their 'study sheds some light on where attention may come from and how attention level may be affected (…), it certainly leaves us with even more questions for future agenda-setting research' (p.415). Liu et al.'s study, with its sophisticated longitudinal design, then provides a salutary reminder of the challenges of mapping the highly complex dynamics and interactions of media, public, political and scientific agendas or meaning-creating forums.

In their exemplary review of agenda-setting research on environmental issues, Trumbo and Kim (2015) conclude that this perspective's major contribution has been to our understanding of the dynamics of media and public agendas in the longer-term and interactive cycles of attention to the environment. Environmental issues are, as they rightly note, 'embedded in multiple agenda domains; not just policy, media, and public, but blogs, advertising and public relations campaigns, efforts by non-governmental organizations, websites, and many others' (p.322) and it is through mapping of the dynamic interactions of these agendas that we can begin to get a picture of how environmental debate is constructed, manipulated and driven forward in public communication.

Quantity of coverage theory (QCT)

Most environmental issues are characterised, at least in their early stages of becoming defined as issues for public and political concern, by considerable degrees of scientific uncertainty and disagreement, as well as by public controversy about appropriate ways to deal with them. In this context, a particularly tantalising model of media influence is the 'quantity of coverage' model suggested in the early 1980s by American sociologist Alan Mazur. In a seminal – and much quoted – article published in the *Journal of Communication* in 1981, Mazur suggested that increased media coverage of scientific controversies seemed to have at least one simple influence on public attitudes: 'When media coverage of a controversy increases, public opposition to the technology in question (as measured by public opinion polls) increases; when media coverage wanes, public opposition falls off' (Mazur, 1981: 109).

This was not only a beautifully simple but also quite a brave statement to make at a time when the tide of media and communication research had long since (i.e. since the 1940s and 1950s) turned emphatically against any notions of such simple cause–effect relationships between media coverage and public opinion. Interestingly, and this perhaps explains some of its general appeal within the research community, Mazur's suggestion was not far removed from the agenda-setting hypothesis (which also, as we have seen, focuses on general rather than specific influences, on 'what people think *about*' rather than on 'what they *think*') and, indeed, relied on the very same methods and type of data, that is, content analysis of media coverage and surveys of public opinion. But where agenda-setting research has developed in the direction of ever-further differentiation of media content and the particular framing of media messages (McCombs et al., 2011; McCombs, 2014), Mazur takes the agenda-setting hypothesis in the opposite direction by suggesting that it is the sheer quantity and intensity of media coverage, not the balance of positive and negative messages within such coverage, that essentially determines public reaction, and that increased coverage (whether positive or negative) leads to increased public opposition:

> My thesis is that the amount of reporting about an environmental or technological hazard, rather than what is reported about the topic, is the primary vehicle of communication about such risks, and that the beliefs of the audience follow directly from the intensity and volume of reporting.
>
> (Mazur, 1990: 295).

In addition to the evidence provided by Mazur's own studies of the relationship between media reporting of nuclear power/waste, fluoridation, chemical waste, avian flu, climate change, fracking, and so on, and public opinion, a number of other studies have provided further support. Wiegman et al. (1989), in a study of Dutch newspaper reporting and reader reactions regarding technological and environmental hazards, found that increased exposure to media coverage correlated with negative public reactions. Frewer (2002), in a study of media reporting and public risk perception in relation to genetically modified foods, found that increased media reporting led to increased public perceptions of risk and related negative consequences. When media reporting declined, public perceptions of risk likewise reduced. Studies of media reporting and public attitudes to biotechnology have pointed in a similar direction, although not as strongly (Gaskell et al., 1999; Gutteling, 2005).

In more recent work, Mazur (2016: 209) summarises the key propositions of quantity of coverage theory as follows:

1. People do not usually attend to the detailed content of news coverage; instead they absorb simple images of hazards [...].
2. People are affected more by the quantity of coverage, especially the repetition of simple images, than by detailed content [...].
3. Public concern about a hazard, oppositional activity, and precautionary action by government rise and fall with the quantity and saliency of news coverage.
4. The quantity of coverage given an alleged hazard is determined more by 'externals'—such as the prominence in the news of related issues, and relationships among journalists and their sources—than by

authoritative evaluations of the scientific validity or severity of the hazard.

5. Most environmental risk stories of national or international scope are first brought to widespread attention by a small, central group of large news organizations including major newspapers (e.g. *The New York Times*), wire services (Associated Press is the largest), and television networks, and by prominent sources including government and environmental spokespeople [...].
6. Therefore, the rise and fall of widespread public concern and governmental action may be traced back to the rise and fall of coverage by the central media.
7. Risk issues given high coverage by major American news organs are often picked up by news media of other nations, more so than American news organs pick up issues first raised elsewhere.

Mazur (2016) uses quantity of coverage theory as a framework for understanding media and public opinion dynamics in relation to emerging controversy about shale gas extraction using the technique of hydraulic fracturing, or 'fracking' as it is now commonly referred to. Comparing trends in mainstream media coverage and evidence from a number of surveys of public opinion regarding fracking, Mazur – drawing from a variety of evidence including interviews with key reporters and stakeholders in the fracking controversy – succeeds in piecing together a persuasive account of key factors in the dynamics and development of public controversy about fracking. In addition to interesting observations on catalysts of media coverage – in this case, how the Deepwater Horizon (offshore) accident and oil spill was a significant catalyst for a major *New York Times* series of articles about the (onshore) use of fracking for gas extraction, and the role and career of the documentary film *Gasland* – Mazur's study offers insights into intermedia agenda-setting dynamics and hints also at the importance of evocative images, in this case the *Gasland* documentary image of a homeowner igniting water (polluted by the fracking process) flowing from his kitchen tap. While recognising the complexity of factors in play in driving forward public communication and controversy about fracking, Mazur's study confirms the simple thesis of the quantity of coverage theory, namely that increased amount of coverage of a controversial technology leads to overall increased public concern and opposition.

The core lesson that Mazur draws from this is worth quoting at length because it resonates well with key arguments in claims-making and framing theory, namely the argument that one of the most powerful influences on public communication may be not the power to place issues on the public agenda for debate, but the power to keep issues from becoming visible or emerging on the public agenda (see the discussion in Chapter 2 and reference to Edelman (1988), Wallack et al. (1999), etc.).

> From a purely tactical perspective, proponents of a risky technology are better served by low levels of news coverage. Conversely, opponents should strive to maximize publicity. The explicit content of news stories is not especially important since few people read extended or technical news stories anyway. What is critical in raising opposition is the frequent repetition of simple images that convey a sense of hazard, or at least a sense of uncertainty. What is critical in defusing opposition is minimizing attention to the controversy.
>
> (Mazur, 2016: 221).

Key propositions of quantity of coverage theory are borne out and confirmed by meta-analyses and reviews. Carmichael and Brulle (2017), in their comprehensive meta-analysis of studies of public opinion on climate change from 2001 to 2013, thus confirm that it is the frequency and prominence of news stories, rather than the balance of evidence presented about a controversial technology or issue, that influences public opinion and particularly leads to 'heightened issue salience', which in turn may mobilise publics to seek out further coverage. Perhaps more importantly, Carmichael and Brulle (2017: 232) also conclude that while 'media coverage exerts an important influence, it is itself largely a function of elite cues and economic factors', and that the communication 'to the public of scientific information on climate change has no effect' on public opinion, but that 'political mobilization by elites and advocacy groups is critical in influencing climate change concern'.

Cultivation analysis

Cultivation research, first articulated and developed by George Gerbner and colleagues in the late 1960s (see, for example, Gerbner et al., 1994), centres on the simple and intuitive hypothesis that the more audiences (and here the original cultivation analyses, as well as most cultivation analyses since, have focused on television viewers) are exposed to media content, the more likely are they to hold beliefs about reality that are consistent with the media's portrayal of reality. The fundamental assumption of the cultivation argument is that for all its apparent diversity, television essentially offers a relatively consistent and repetitive set of narratives, images and values, and that, over time, viewers come to see these as the dominant narratives, images and values of society, as 'reality'.

The application of cultivation analysis to media and environmental issues was first pioneered by Shanahan in the early 1990s (Shanahan, 1993) and has since been applied by Shanahan and colleagues in numerous studies. One problem for the use of cultivation research in relation to environmental issues is that, rather than systematic over-representation – as in the case of depictions of violence, crime and law enforcement – the 'environment' has not generally been a prominent focus in television entertainment, leading to what Shanahan (1993) calls 'cultivation in reverse'.

> Using a cultivation approach, Shanahan, McComas and their collaborators have found that heavy viewers of television viewing are less likely to be environmentally concerned and less willing to pay more (either through taxes, prices, or a lower standard of living) for the environment, although only in the absence of controls. Heavy television viewers also displayed lower levels of trust in science and technology and less environmental knowledge than lighter viewers of less television: relationships that generally withstood controls. [...] (Shanahan, 1993; Shanahan and McComas, 1997, 1999; Shanahan et al., 1997). Holbert, Kwak, and Shah (2003) recently showed a positive relationship between environmental concern and attention to television news and nature documentaries while also finding no evidence of a relationship between environmental concern and three different types of attention to entertainment television.
>
> (Besley & Shanahan, 2004: 864)

In a 2015 update and overview of whether and how television cultivates environmental concern, Shanahan and his colleagues note that while research has not produced strong evidence of a cultivation effect, key arguments about television viewing and environmental concern remain valid. These are: 1) that more time spent watching television means less time experiencing nature or the outdoor environment; 2) that the environment as a social problem remains comparatively invisible in television entertainment programming; and 3) television's focus on materialism runs counter, in the words of Jennifer Good (2009, 2013), to environmental concern. Perhaps one of the key challenges for the cultivation perspective is that its premise of a mass-mediated flow of relatively uniform entertainment fare, characterised by clear trends of dominant messages about the environment, no longer fits either the increasing diversity of media forms, channels and images or the increasingly diverse ways in which publics encounter and consume mediated representations of the environment.

Box 7.4

A different kind of 'effect': the use/construction of mediated communication itself as an actor in public debate

Like public opinion, media coverage itself often becomes a central referent and actor in the rhetoric of public controversy. Selective use of media stories as 'evidence' is both tempting and easy for the simple reason that the media are so highly visible and therefore an easily identifiable reference point. In the same way as 'public opinion' or 'what the public wants' often get invoked rhetorically as a way of lending legitimacy to particular arguments, so too will politicians, scientists, experts and other key players in environmental debate, often point to the media and – carefully selected – media news stories either to back up their arguments or, perhaps more often, to blame the media for misinforming the public and for stirring up public panics, hysteria and rash political decision-making on important issues or social problems.

The age-old rhetorical strategy is essentially to shift the focus of debate away from the balance of (scientific) evidence onto the messenger (the media) and the way in which different positions on a controversial topic are given prominence and reported impartially, 'objectively' or 'accurately'. The assumed significant influence of 'inaccurate' or 'imbalanced' reporting on public understanding and political decision-making is behind a large historical body of communication research on accuracy and objectivity in news reporting of controversial science, technologies and issues (Hansen, 2016), from – in environmental communication – nuclear power, agricultural and food biotechnology to climate change.

A now classic example of enlisting 'mediated communication' as a key actor (and problem) in debate about public understanding and political decision-making on climate change is Al Gore's film (and book) *An Inconvenient Truth – a Global Warming* (Guggenheim & Gore, 2006).

With analogies drawn to media reporting on the long-running 'controversy' about the extent of harm caused by and efforts to restrict tobacco/smoking, Al Gore – at a relatively late point in his film, that is, scene 26 of 32 – makes the argument that the media grossly misrepresent the scientific consensus on climate change, and he goes further to say that the public misconception about the causes of global warming has been 'deliberately created by a relatively small group of people' (note the careful avoidance of any specific identification).

Gore invokes references to advertising and media news coverage, and makes the simple but effective comparison of scientific opinion on climate change as reported by: 1) articles/research published in scientific peer-reviewed journals, and 2) by newspapers. Speaking to an audience in a subtly lit auditorium, Gore delivers his narrative from a podium against the background of a large central screen and a much smaller screen on the left. The accompanying PowerPoint show is projected onto both screens. With rows of little figures symbolising the scientist/expert authors of the 928 journal articles reviewed rapidly decreasing, and numbers rapidly dropping from 928 to 0, Gore makes his point that none of the studies disagreed with the consensus that 'greenhouse gas pollution has caused most of the warming of the last 50 years'. He then goes on to refer to a study of media coverage over 'the last fourteen years' showing that 53% of newspaper articles implied that there was doubt about the cause of global warming. The point of Gore's reference to media coverage is to indicate that arguments about the causes of global warming are deliberately being manipulated and that in view of the nature of media coverage, as illustrated by his example, it is 'no wonder that people are confused'.

While the focus and analysis, in *An Inconvenient Truth*, of media reporting as a key influence and actor in public debate is effectively executed and indeed draws on directly relevant scientific research for its argument, mediated communication is much more commonly invoked as evidence or a key actor in a loose and unsubstantiated way that simply exploits the high public visibility of media reporting, the sense of familiarity with 'what's in the news', and the power of selective anecdotal examples.

In the first edition, I drew attention to prominent climate change sceptic Bjørn Lomborg's (2007) skilful rhetorical strategy of continuously invoking the news media's reporting on climate change as a key actor and cause of public/political confusion or misunderstanding regarding climate change. The crux of the strategy is to use 'media reporting' in a general and impressionistic sense as evidence of misinformation, bias, scaremongering and public misunderstanding. Occasionally, this may be augmented with reference to actual communication research evidence (such as systematic content analyses of media reporting), carefully and selectively chosen to support one's key argument, as in the case of Bjørn Lomborg, of questioning climate change or political responses to climate change.

The rhetorical use or construction of media reporting/mediated communication regarding climate change as a key and influential (on public opinion/political decision-making) actors is of course not confined to particular stances or sides of the debate, such as that of climate sceptics. In the late 1990s, Cottle (1998), for example, offered a critical analysis of Ulrich Beck's rather generalised referencing of the media and media roles in *Risk Society* (1992). More recently, communications scholar Robert Hackett (2015) has drawn attention to the similarly general referencing of the influence of the media on responses to climate change, in Naomi Klein's (2014) *This Changes Everything: Capitalism vs. the Climate.*

Framing

While cultivation analysis proper focuses on the general, long-term and consistent trends and messages of media content, research on narratives and framing in media depictions of environmental issues has emphasised how individual (news or entertainment) stories are structured to, as it were, produce a particular response or conclusion in the minds of viewers.

Framing research in particular has drawn attention to how the principles of 'selection' and 'salience' (Entman, 1993; Nisbet & Newman, 2015) in media content help structure audience responses by directing attention to: 1) what the issue/problem is; 2) who/what is responsible; and 3) what is to be done about the issue/problem, that is, what the solution is (Ryan, 1991).

Drawing on the framing categories developed by Gamson and Modigliani (1989) in their analysis of nuclear issues and popular culture, studies of media and environmental issues have demonstrated how key interpretative packages ('progress', 'economic prospects', 'ethical/moral', 'Pandora's box/runaway science', 'nature/nurture', 'public accountability', etc.) are strategically deployed and manipulated by sources in public environmental debate, with significant potential implications for both the nature of media coverage and the mobilisation of public understanding, opinion and behaviour with regard to controversial issues such as nuclear energy, biotechnology and climate change (Nisbet & Newman, 2015).

It is important to note that frames – and this indeed helps explain their potential power – are not simple indications of binary stances such as positive versus negative or pro versus anti, but rather ways of drawing attention to and foregrounding particular discourses or interpretations over others. As Nisbet and Newman (2015) remind us:

> [F]rames serve as general organizing devices for public debate and should not be confused with specific policy positions. In other words, each frame can relate to pro, anti, and neutral arguments, though one set of advocates might more commonly activate one cultural schema over others.
>
> (324)

The often-observed tendency of news reporting (due largely to the journalistic value of objectivity) to balance opposing views in reporting of controversial environmental issues, from nuclear power and genetically modified crops to climate change, has the effect of influencing and perpetuating public perception of widespread scientific uncertainty long after scientific – and even in some cases, political – consensus has been reached. Studies of climate change news in particular have shown the influence of uncertainty framing, and on whether news is framed as political or science news, on public understanding and perception. An early experimental study by Corbett and Durfee (2004) found that people exposed to opposing views in news reports were more likely to be uncertain or express doubts about climate change.

More recent studies have confirmed how political news frames versus science news frames impact on public understanding and attitudes with regard to climate change. These studies (Hart et al., 2015; Nisbet et al., 2015; Zhao et al., 2011) show that the framing of climate change in political news (and audience attention to political news) makes little or no contribution to public understanding of climate change, or indeed is negatively related in the sense of reinforcing doubt about climate change. Conversely, these studies show that the framing of climate change in science news increases and reinforces both public knowledge about climate change and beliefs that climate change is happening and needs addressing.

Particularly promising indications of the significance of framing comes from work by Maibach, Nisbet and colleagues on how reframing communication about climate change in terms of local – as opposed to global – concerns and in terms of a health frame or a security frame impacts on public understanding and, particularly, on public engagement:

> Framing climate change in terms of public health stresses climate change's potential to increase the incidence of infectious diseases, asthma, allergies, heat stroke, and other salient health problems, especially among the most vulnerable populations: the elderly and children. In the process, the public health frame makes climate change personally relevant to new audiences by connecting the issue to health problems that are already familiar and perceived as important. The frame also shifts the geographic location of impacts, replacing visuals of remote Arctic regions, animals, and peoples with more socially proximate neighbors and places across local communities and cities. Coverage at local television news outlets and specialized urban media is also generated.
>
> (Nisbet & Newman, 2015: 330)

Key to understanding variations in how framing influences public understanding and engagement with mediated environmental communication is also the recognition that different frames resonate differently with different publics or 'interpretative communities'. As Nisbet and Newman (2015: 337) explain, these are groups of 'individuals who share common risk perceptions about climate change, reflect shared schema, mental models, and frames of reference, and hold a common sociodemographic background'. While the notion of differentiated publics has been recognised in communication research since the mid-20th century, what is new and different in the present century is the increasingly fragmented and diverse digital media environment, where publics can and do increasingly attend to and select the type of media and news content which confirm and reinforce, rather than change or broaden, existing views and beliefs characteristic of one's interpretative community (see also Roser-Renouf et al., 2015).

An argument and observation that is common to framing theory and several of the perspectives discussed earlier in this chapter is the notion that public understanding of general social issues, such as climate change and other environmental issues, relies on and is influenced by general cues rather than detailed study. That is, on most issues that we are not directly studying, experiencing or connecting with, we form our understanding, perception and opinion on – sometimes fleeting – observations of key cues from our symbolic environment, notably as we are aware of it from the mediated content that we consume:

> With limited time and ability to process complex information, as we move through our daily lives trying to make sense of a constant flow of ambiguous signals, situations and choices, we are heavily dependent on shifting cues that set the context for our perceptions. In this regard, both as a communication necessity and as a persuasion strategy, when experts, advocates or journalists 'frame' a complex environmental issue, they differentially emphasize specific cues relative to that complex issue, endowing certain dimensions with greater apparent relevance than they would have under an alternative frame.
>
> (Nisbet & Newman, 2015: 229)

Framing theory then adds to our understanding of public-mediated environmental communication and its influences by focusing on selection and salience in public communication, and by demonstrating how variations in prominence and absence (selection) and in context/perspective/framework/discourse (e.g. science, health, security, politics, religion, etc.) influences how we take our cues from and make sense of public-mediated communication about the environment.

While studies of media coverage and public opinion have amply demonstrated the difficulties and complexities of mapping their interaction, there is little doubt that the media serve as an important public reservoir of readily available images, meanings and definitions about the environment. The media are an important public arena (Hilgartner & Bosk, 1988), where different images and definitions – 'sponsored' by different agents, groups and interested parties – compete and struggle with each other. Environmental meanings, messages and definitions communicated in any one single medium, format or genre are unlikely to exert a simple linear influence on public beliefs, understanding or behaviour. Not only do different publics attend to different media and to different extents, but it is also clear that there is great variation in how different publics use and make sense of the information and messages that fill our public sphere (Whitmarsh, 2015).

While the notion of differentiated publics engaging in differentiated ways with media and public communication has been understood since the very early days of mass communication research of the mid-20th century, it is only in this century that we have witnessed real advances in its application to environmental communication. As we have seen, research on climate change and related environmental issues in the first decades of the present century has provided real advances in our understanding of the significance of political outlook, politicisation of debate, and interpretive communities in terms of not just what mediated content (from the level of different media to the level of narrative genres and formats) different publics attend to, but in terms of how they engage with, use or ignore, and make sense of what is being communicated.

The media and public communication, in their broad and diverse totality, provide an important cultural context from which publics draw both vocabularies and frames of understanding for making sense of the environment generally, and of claims about controversial environmental issues more specifically. Capturing the dynamic nature of what Burgess and Harrison (1993) referred to as the 'circulation of claims in the cultural politics of environmental change' continues to be one of the foremost challenges for environmental communication research. But it is also a challenge on which very significant progress has been made in the first two decades of the present century, particularly through research such as agenda-setting and framing analysis which continues to map the intricate dynamics of claims-making, mediated communication and analysis of how different publics and policy-makers make sense of, react to, create meaning and take action (or not) with regard to public media and communications messages about the environment.

Given the profusion of media images, the diversity of media, and most particularly the diversity of public consumption of media images, the relationship between media images of environmental issues and public perceptions, attitudes or understanding regarding environmental problems is clearly then one that is not best addressed by asking the simple question 'what are the effects of the media?' Nor is the role of the media best addressed in terms of one-directional linear models of communication effects which assume that media reporting *cause* public or political opinion – or vice versa. To begin to understand the role of media in the communication of environmental issues, we need a different vocabulary and visualisation. We need to think in terms of the public 'circulation of claims' about the environment and we need a vocabulary with terms like dialectic, interaction, reinforcement, engagement, information loops, neural networks, multidirectional, resonance and parallel forums of meaning creation.

Conclusions

Media coverage and mediated representations of environmental issues matter: they play a prominent and significant role in the social construction of environmental issues – not least those that we have little or no direct experiential access to – although pinpointing and quantifying the exact nature and extent of media influence remains a challenging task for communication research. However, more often than not, this is because we are asking the wrong questions. Questions that are themselves stuck in simplistic and outmoded assumptions about direct simple linear media 'effects' are likely to only generate answers that – while statistically often highly robust and sound – tell us little to help explain the truly complex dynamics of public and political opinion formation on environmental and risk issues.

Mediated communication (like 'public opinion') is itself an important reference point in the public and political construction of social controversies, and an important chess piece in the public discursive game-playing characteristic of the rise and fall of major social issues in the public arena.

Media coverage impacts on and interacts with public opinion and political opinion/policy-making in complex ways that are more likely to fit the visual images/metaphors of loops and spirals than those of the one-directional arrows of direct linear effects models of communication.

The media, media representations and mediated communication generally of environmental issues are best conceived of as a continuously changing cultural reservoir of images, meanings and definitions, on which different publics will draw for the purposes of articulating, making sense of, and understanding environmental problems and the politics of environmental issues.

Environmental issues don't simply present themselves as issues for public and political concern. Environmental issues – and public concern about the environment – are socially constructed. They become issues for public and political concern through complex dialectical processes of claims-making and counter-claims-making. As we have seen, communication and media are core to these processes. The media are at once a public arena where claims-makers compete to have their claims heard and publicised and are themselves an active influence on the selection and framing of claims-makers' claims.

It is to the analysis of the dynamic interaction of media, publics, politics, claims-makers and social institutions that we must turn if we wish to begin to understand the processes by which some environmental issues are defined as social problems while others remain socially invisible, why some environmental claims succeed in the public sphere while others wither on the vine or fall by the wayside.

Further reading

Part IV of the *Routledge Handbook of Environment and Communication* (Hansen & Cox, 2015) provides state-of-the-art reviews of research – from a range of perspectives and communications models – on the social and political implications of mediated environmental communication. See the following chapters:

Chapter 26: Mapping media's role in environmental thought and action (pp. 301–311) by Susanna Priest.

Chapter 27: Agenda-setting with environmental issues (pp. 312–324) by Craig Trumbo and Se-Jin 'Sage' Kim.

Chapter 28: Framing, the media, and environmental communication (pp. 325–338) by Matthew C. Nisbet and Todd P. Newman.

Chapter 29: Analysing public perceptions, understanding and images of environmental change (pp. 339–353) by Lorraine Whitmarsh.

Chapter 30: Publics, communication campaigns and persuasive communication (pp. 354–367) by Todd Norton and Natalie Grecu.

Chapter 31: Engaging diverse audiences with climate change: message strategies for Global Warming's Six Americas (pp. 368–386) by Connie Roser-Renouf, Neil Stenhouse, Justin Rolfe-Redding, Edward Maibach, and Anthony Leiserowitz.

8 Environment, media and communication – looking back, looking forward

In this concluding chapter, I first offer a condensed overview of some of the main emphases and conclusions arising from the survey in this book of the field of research on environment, media and communication, and then conclude with some suggestions for the way ahead in environmental communication research. Starting with a brief reiteration of the importance of media and communication in the rise and development of environmental concern, I delineate the disciplinary and theoretical context and some of the main emphases of environmental communication research. Key trends are then summarised under the headings of news media representation of the environment, non-news media representation, and environmental communication in the digital media and communications landscape. I conclude with a summary of the major achievements of environmental communication research to date, and – pointing to some of the key challenges ahead – I offer suggestions for where, particularly in light of the rapidly changing nature of the media and communications landscape, future research emphases need to be focused.

The rise of environmental communication research

Research on the mediated communication of the environment has come far since it first started emerging in the 1960s. The field of environmental communication, as it is now generally known, has particularly developed and consolidated in exciting ways in the present century. In the second decade of the 21st century, we have thus – as detailed in Chapter 1 – seen the establishment of the first international association of environmental communication researchers and practitioners (the IECA), the launch of book series devoted to environmental communication, the publication of multiple monographs and textbooks on diverse aspects of media and environmental communication, and the publication of synoptic collections, handbooks and encyclopedias focused on environmental communication (always a promising sign of consolidation and of the field reaching a first stage of maturity).

As I have sought to emphasise throughout, the rise of environmental communication research has gone hand in hand with the rise of the 'environment' as a social and political issue. This is no coincidence, as indeed the public identification and construction of the environment as an 'issue' depends crucially on mediated public communication. The central argument of the book is thus that media and communication have been and continue to be central to the public and political definition of the environment. If we want to understand our relationship, as societies and as individuals, with the environment – including how we use, live in, engage with, think about, exploit, protect and care for the environment – then we need to understand how the environment is 'constructed' through mediated communication. Furthermore, we need to acknowledge that all communication is done for a purpose and that communication about the environment – as with all political issues – is a space of contestation and competing definitions and interests. In this respect, the key question to ask of mediated environmental communication is not whether it is 'accurate' or 'balanced' (or impartial), but rather to establish how and why some definitions become more prominent and successful than others in the public sphere, and – crucially – to determine whose interests are served (and whose disadvantaged) by this.

The establishment and rise of the environment as a focus for social and political concern then has been facilitated by the growth in (mass) mediated forms of communication, including perhaps particularly the growth of visual media. Traditional broadcast and print media and newer forms of digital communication have been central and instrumental in defining 'the environment' as a concept and domain, and in bringing environmental issues and problems to public and political attention. The inherently long timescales and often low immediate visibility of many environmental changes or problems mean that much of what we as publics know or recognise as 'the environment' or more particularly as 'environmental problems', we know or perceive through mainstream media and other mediated forms of communication. And media and forms of communication are subject to multiple pressures, influences and constraints – notably economic and technological, but also factors to do with the (unequally distributed) communicative resources and skills of those who contribute to, participate in, draw on and engage with communication about the environment.

Research on media, communication, public opinion and the environment draws, as we have seen, from a broad range of both humanities and social sciences disciplines, and it owes much of its innovativeness and dynamic development to this productive mix of theoretical and disciplinary traditions.

The development of environmental communication research as a distinctive strand within media and communication studies generally is hardly surprising when considering the centrality of public media and communication processes to drawing public and political attention to 'environmental problems'. Originally focused predominantly around the study of mainstream news journalism, media coverage of the environment and associated public opinion, environmental communication research has benefited from developments in the closely related fields of science communication and risk communication. Significantly, it has expanded from originally relatively narrow concerns with news and journalism to examine a much broader range of media, genres and forms of communication. It has also expanded from a narrow focus on journalism to draw on a much richer body of theories and approaches to help understand and elucidate the broader social, political and cultural roles of environmental communication.

As outlined in Chapter 2 and applied in subsequent chapters, the social constructionist perspective helped move communication research on environmental problems out of journalism studies trapped in circular concerns with balance, bias and objectivity, and proved a productive inspiration for attempts at grappling with sociological interpretations of the media's role in public and political controversy about the environment. Within mainstream media and communication research, organisational and cultural perspectives on news production, agenda-setting research, and, since the 1990s, the concepts of 'framing', 'interpretive packages' and 'cultural resonances' have provided productive – and often overlapping – frameworks for analysing environmental communication, not just in news media but across all genres and forms of mediated communication.

Central to the analysis and understanding of mediated environmental communication, and indeed prominent in the research surveyed in this book, is the close attention to language, rhetoric, visualisation and discourse in public communication about the environment. There is thus a clear recognition that these dimensions are central components of how issues are constructed and 'framed' and how, in turn, particular messages/meanings are conveyed and boundaries set for public understanding, engagement and public interpretation/opinion regarding environmental issues.

Since the 1960s, the 'environment' has become firmly established on the public and political agenda, and as argued throughout this book, media and communication processes have been central to the public definition of the environment as a field for public and political concern. Given the centrality of mediated communication to the social, political and cultural construction of the environment, it is not surprising that the principal focus and point of departure for research has often been on *media representations* of the environment. It is also the case, however, that the analysis of media representation is almost invariably done with a view to understanding both the processes of its production and its wider social and political implications.

News media representation of the environment

Easily the most prolific focus of research on mediated environmental communication has been and continues to be that of news coverage of environmental issues and controversies. Such research has contributed considerably to our understanding of why some environmental issues are successfully constructed as issues for public concern, while others quickly vanish from the media agenda and from public view. Studies of news reporting are thus an invaluable starting point for discovering the multiple factors that influence the extent and nature of media coverage – and indeed whether an issue makes it onto the media agenda at all. They provide a point of departure for mapping and understanding the media career of environmental issues and the seemingly cyclical nature of environmental coverage and public environmental concern in what Downs (1972) presciently referred to as 'the issue-attention cycle'.

Studies of news coverage of the environment have drawn on key developments in the sociology of news over the last half-century to provide insights into how such coverage is influenced by multiple factors relating – as surveyed particularly in Chapters 3 and 4 – to journalistic practices, news values, organisational arrangements within the media, ownership and editorial control in news media, as well as the wider cultural context in which claims are made about the environment. Early research in the 1970s showed that the extent and nature of news coverage of the then relatively novel notion of 'the environment' often depended closely on whether news organisations had a dedicated environment beat with specialist environmental journalists. It also showed the significant power of sources to influence the extent and nature of news definitions of environmental issues.

The changing relationship between sources and journalists, and the role of sources in influencing the agenda and nature of news coverage of the environment, have continued to be central foci for research. Such research has mapped how significant technological and economic changes have impacted on media organisations and on environmental journalism. Notably, the rise of new digital media and forms of communication, increasing competition, and economic pressures have combined to reduce or eliminate traditional environmental beats and journalistic roles within traditional news organisations. The contraction in environmental journalism has, as argued in Chapter 4, been matched by an expansion in sources' use of public relations strategies, resulting in an overall significant shift of power from journalists to sources in terms of ability to influence the agenda and nature of public debate about the environment.

Snapshot analyses of news coverage of environmental events, disasters or environmental issues have been highly effective in demonstrating the operation of particular news values. They have shown the persistent 'authority orientation' of news coverage; that is, news media tend to turn to politicians, scientists, experts and establishment representatives for definitions of issues, rather than to NGO or environmental pressure group representatives or indeed to 'victims' or other members of the general public. And they have demonstrated the thematic emphases and framing characteristic of environmental issues coverage, for example a tendency to represent environmental problems as distinct and isolated occurrences rather than placing them in their wider interconnected context.

Longitudinal studies of environmental news coverage have provided insights into the complex dynamics and careers of environmental issues, throwing light on what 'drives' environmental coverage and on how different public agendas (news media, politics, science, public opinion, etc.) interact. Mapping the significant fluctuations over time in media attention to climate change and other environmental issues, longitudinal research has greatly facilitated and enhanced recognition of the key roles of claims-making practices, news values, journalistic practices, media organisational routines, and interaction with other issues in determining the extent and framing of coverage.

Longitudinal studies of media coverage of the environment have contributed significantly to a more nuanced understanding of how media agendas are constructed and of how they interact with other agendas as demonstrated, for example, by agenda-setting research. Mediated representations and communication of the environment are a particularly good quarry for agenda-setting research, as environmental issues are generally both relatively unobtrusive and slow-developing, meaning that we as publics are more likely to learn about them through media than through direct observation or experience.

One of the most interesting and productive developments in research on media representation of the environment has been the increasing use of the concept of framing. Framing analysis examines the principles of selection and emphasis/salience in media content, which may contribute to the structuring of public and political responses by directing attention to how an issue or problem is defined, who is cast as being responsible, and what solutions are proposed or implied for remedying the issue/problem. Like agenda-setting research, framing analysis provides a productive framework for connecting the production, content and interpretation of mediated environmental communication. Thus, this type of analysis has helped demonstrate how frames are strategically deployed and manipulated in public environmental debate, with significant potential implications for both the nature of media representation and the mobilisation of public understanding and engagement.

Comparative research – comparing media representation of the environment across time and/or across different countries or cultures – has yielded particularly interesting insights into how mediated environmental communication is further circumscribed by cultural, national and regional differences impinging, for example, on journalistic values and practices. Just as the comparison between countries or cultures can provide important clues to how the extent and nature of environmental issues coverage, and its framing and inflection, is circumscribed by cultural resonances and shaped by journalistic traditions and organisational constraints, so too is it important to recognise how different types of media and different media formats also afford very different possibilities in terms of what is/can be communicated about the environment. Studies have thus shown the significant differences between national and local/regional media in how they frame environmental issues coverage, how critical/uncritical the reporting is and what types of sources predominate.

Non-news media representation of the environment

While the news media representations of the environment have historically been by far the most prominent focus for research, an important body of research on other types of media and genres has begun to emerge. There is thus now a considerable and expanding body of work on the representations of nature, the environment and environmentalism in film (including animated film and cartoons), advertising, and in documentary and entertainment television. As argued particularly in Chapters 5 and 6, broadening the focus of research well beyond the news media is important in order to understand how public understanding and policy with regard to the environment draw from and interact with not just news representations but also the messages, images and ideologies about the environment that circulate in the wider cultural and symbolic environment.

Studies of film and other non-news communication formats have begun to demonstrate how these more flexible formats offer opportunities for more nuanced and varied, potentially more progressive and perhaps problematising, perspectives on key environmental issues, such as climate change, the use of natural resources, our relationship with nature, sustainability, and so on. A comprehensive body of research built up since the early 1990s on television entertainment representations of the environment has tended to conclude that the environment has had a comparatively low profile in such media fare.

Studies of advertising representations have shown that while explicitly 'green' or 'environmental' advertising comes and goes, uses of – often romanticised or nostalgic views of – nature and appeals to the natural tend to feature prominently and relatively consistently over time in the marketing of a wide range of products and ideas. Advertising representations draw on – and rework for the purpose of selling products and ideas – deep-seated cultural understandings of nature, the natural and the environment. In doing so, advertising images of nature and the environment contribute – like other media representation – to wider public definitions and understandings of how we relate to, consume, protect, sustain and use the natural environment.

Environmental communication in the digital media and communications landscape

Increasingly, the traditional distinction between news media and non-news media content, and between content genres such as advertising, drama and fictional entertainment, documentary, and so on, is becoming less clear-cut and more fluid, as media and communications technologies converge. The new digital media environment, characterised by the increasing collapsing of time and spatial boundaries, and by the rise of online and social media, has far-reaching implications for environmental communication.

Environmental campaigners and pressure groups have been quick to seize the greatly enhanced opportunities afforded by new digital and social media in terms of exploiting or bypassing the control over information dissemination traditionally exercised by established media organisations, but also as a way of mobilising public support and protest at local, national and international level. This is not a case of new media campaign strategies replacing more traditional approaches to influencing the media and public agenda, but rather a matter of complementing, augmenting and enhancing traditional campaigning and mobilisation strategies with the access and collapsing of time and geographical boundaries afforded by new media.

While environmental campaigners and pressure groups may have been amongst the first to seize the opportunities offered by networked digital communications technologies and new social media, governments, corporations and big businesses have not been slow to likewise exploit these opportunities to engage with and influence public debate. This raises important challenges for environmental communication research in terms of analysing and mapping the rapidly changing media and communications environment, and the implications for shifting balances of power and control in the dynamics and politics of environmental communication.

Conclusions and suggestions for the way ahead

Environmental communication research has come a long way in the last few decades. It has consolidated itself as a distinct multidisciplinary field of inquiry, drawing productively from both the humanities and social sciences, and it has demonstrated the advantages of bringing diverse theoretical frameworks and analytical approaches to bear on the analysis, not just of news media but a much wider range of media and communications forms and processes.

As I have noted elsewhere (Hansen, 2015b) the main achievement is perhaps the considerable advances in the last two decades towards an increasingly sophisticated understanding of the complex processes involved in the social 'construction' of the environment as an issue for public concern. We thus now know a great deal about the news management, publicity and campaigning practices of environmental claims-makers in the public sphere, about environmental journalists and environmental journalism, about the organisational and economic pressures impinging on media organisations and their handling of the environment, and about the social, political and cultural implications of communication about the environment.

Research on the communication strategies of sources and on source–journalist relationships has made important advances in terms of showing how successful claims-making in the public sphere is closely related to the (economic and organisational) resources and political power commanded by key claims-makers. Much media content research has likewise referenced the ideological nature and implications of media coverage, and indeed alluded to the inequalities and imbalances of power in public communication demonstrated through the persistent imbalances in source accessing and the framing of different sources.

There is, however, a need to push this frame of enquiry significantly further both in terms of uncovering the deeply ideological nature of public communication, but more particularly, I would argue, in terms of uncovering how communicative 'power' in society is profoundly unequally distributed. The task in this respect is to reconnect, empirically and within clearly articulated theories and models of communication processes, the study of the production, the content and the social/political implications of environmental communication. And further, to demonstrate, building on the insights already gained from research in each of these domains, how media and 'mediated' communication about the environment and environmental controversy are invariably propelled/

influenced and manipulated by competing agencies and interests, which in turn have very different degrees of power and very different communicative resources at their disposal.

In approaching and reconnecting research on the three domains of environmental communication research (Hansen, 2011), key recognition needs to be given to the rapidly changing nature of both *media* and *communication*. As we have seen, the nature of media organisations has changed significantly during the present century, and with it so has the nature of work of environmental journalists and others involved in communicating about the environment. All of these changes have had profound impact on the balance of power and the relationship between sources and communicators, with – in turn – implications for how and by whom environmental issues/problems are articulated and defined, who is seen and publicly labelled as responsible or to blame, and what solutions are being or can be recommended.

But perhaps the most profound change of all relates to the technological changes and affordances of the digital communications environment and its implications across all three domains of the communication process. Changes in public communication and in the way we communicate have thus potentially exacerbated trends, such as the documented decline since the 1960s in public trust in science, politics and many other institutions, and how we as citizens engage with information. Thus we have seen the resurgence of the age-old concerns with 'accuracy' and 'misinformation' in news, much of it now under the term of 'fake news'.

Perhaps more significantly, there are profound changes underway in how we engage with news and information about the environment, and, concomitantly, in how communication about the environment is, can be and will be manipulated. The digital communications environment provides us with unprecedented and easy access to virtually any kind of information, 'evidence' and opinion, which is surely a positive thing. But at the same time many of the traditional (and up to a point 'trusted') channels of communication (such as major news media organisations) have declined (although as we have seen in Chapter 7, perhaps not as much as expected) in importance as people's main sources of information. Similarly, many of the traditional mechanisms of 'gate keeping' and 'fact checking' (including those traditionally performed by environmental journalists and other news media professionals) and ways of establishing the credibility and validity of information have been eroded or vanquished altogether.

Lewandowsky and his colleagues (2017: 353), in a thoughtful examination of these trends and how to counter them, thus poignantly argue that what is now often referred to as the 'post-truth' world 'emerged as a result of societal mega-trends such as a decline in social capital, growing economic inequality, increased polarization, declining trust in science, and an increasingly fractionated media landscape'. A key characteristic of this new media and communications landscape, and new forms of communication, is the ability to choose (through custom-designed 'filter bubbles' and 'echo chambers') exposure only or mainly to information which conforms with and reinforces our pre-existing attitudes.

Another characteristic is the change – partly perhaps driven by the changing format/type of communication in the form of short messages, for example, Twitter – in length, tone and rhetorical structure of communication, including the erosion or outright dismissal of evidence-based argumentation, and the concomitant rise of a public political discourse where complex public issues, problems, policies and decisions are reduced to brief 'sound bites' (rhetorically designed for maximum impact, using popular buzzwords, metaphors and slogans, rather than with a view to conveying information) and to what Lewandowsky et al. (2017: 359–360) characterise as a discourse of 'extreme incivility':

> [O]nline political discourse has become characterized by extreme incivility. It has been suggested that Twitter, in particular, promotes public discourse that is 'simple, impetuous, and frequently denigrating and dehumanizing,' and that 'fosters farce and fanaticism, and contributes to callousness and contempt' (Ott, 2017, p. 60). [...] One aspect of this incivility is outrage, characterized as 'political discourse involving efforts to provoke visceral responses (e.g., anger, righteousness, fear, moral indignation) from the audience through the use of overgeneralizations, sensationalism, misleading or patently inaccurate information, ad hominem attacks, and partial truths about opponents'.
>
> (Sobieraj & Berry, 2011: 20)

On the basis of the trends and changes in environmental communication and in environmental communication research surveyed and discussed in this book, I finish (expanding on my suggestions in Hansen, 2016) with the following suggestions for environmental communication research foci and questions as we move forward in this exciting field of study and inquiry:

1. The implications of the rapidly changing communications environment (digital media, social media, user-generated content, civic science journalism, and so on) where a combination of rapid technological development and increasing economic pressures on media institutions result in a much-changed type of mediation. Much of the authority and trustworthiness of traditional media organisations and modes of communication are replaced by a multitude of communications/sources/channels whose motives and handling of science-based evidence become increasingly difficult to assess. How do these changes impact on trustworthiness, credibility and science-/evidence-based argumentation and decision-making?
2. The manipulation of the mediated environment is becoming increasingly complex and diverse. Research needs to catch up with some of the fascinatingly skilful ways in which claims-makers of all types – but perhaps particularly those, such as governments and large corporations, with access to large communication resources – are taking advantage of the affordances of the digital media landscape and influencing public communication about the environment. Some of the key challenges for environmental communication research concern the changing nature of how we interact with, assess and consume information. This involves moving from the focus on traditional media and genres, to examining how claims-makers of all types are increasingly making use of a wide variety of digital/online media and media forms (including gaming) for promoting products, ideas, values and views regarding the environment and our relationship with it.
3. Media and communications are increasingly multimodal – verbal, visual, aural – and research on environmental communication needs to attend not just to the textual (traditionally, the predominant emphasis) or visual communication of the environment, but to the full combined multimodal nature of communication. The strengths and insights of traditional approaches to the analysis of mediated communication need to be mobilised into a multimodal design that considers how the 'meaning' and ideological construction of the environment is influenced through the multiple sign systems of digital communication media.
4. As the media and communications environment changes and becomes increasingly diverse and differentiated, there is an urgent need for a much greater use of comparative research. While promising comparative studies continue to emerge, the predominant mode of research on environmental communication has for a long time comprised mainly studies focusing on a single traditional medium

(most often newspapers, less frequently television and rarely radio) and on news journalism over relatively narrowly defined periods of time. Only by conducting comparative research in the full meaning of this term – that is, comparative across media, across genres, across cultural/political environments and most crucially across time – can we begin to understand the complexity of factors influencing the production, mediation and public engagement with communication about the environment.

5. Just as research is needed on the increasingly active production/promotion and strategic use of environmental communication in the public sphere by multiple sources/communicators (government, NGOs, business, corporations, universities, science institutions, environmental pressure groups, etc.), so too is research needed on how different publics interact with and act upon public information, debate and controversy regarding the environment. More research is needed on how diverse publics interpret and engage with environmental communication, how different publics: a) choose their information channels/media and sources, and b) discern credibility and validity of information in an increasingly diverse media environment replete with 'echo chambers', 'filter bubbles', and a changed mode of non-evidence-based argumentation and rhetorical style.

While research on environment, media and communication has, as we have seen, been around for many decades, there has never been a better or more exciting time than the present to study environmental communication. As the media and communications landscape and the nature of public communication are evolving and changing rapidly, and as public and political views on the environment and environmental policy experience major upheavals and changes (likely not unrelatedly), there is a greater need than ever to understand the roles of mediated public communication about the environment.

This calls for environmental communication that needs to combine the analysis of the three key domains of media and communication: analysis of *content* with analysis of its *production* and its *reception/consumption*. This is not simply a question of bringing together or comparing findings from individual and separate studies from each of these sites/domains, but rather a question of formulating research designs within theoretical frameworks that articulate the relationship and dynamics between the sites.

'Reconnecting' the three domains of research is thus not only a matter of strengthening our understanding of the dynamics which drive and impact on the processes of public communication and the circulation of claims, but it is also a matter of showing how economic, political and cultural power significantly affects the ability to participate in and influence the nature of public 'mediated' communication about the environment.

If we acknowledge that studying and understanding mediated environmental communication in the public sphere is about understanding the processes by which 'truths', 'facts', and 'evidence' about the environment are socially 'constructed', then the key task for environmental communication research should be to establish and understand how and why some definitions/views and 'solutions' regarding the environment become more prominent and successful than others in the public sphere, and – crucially – to determine whose interests are served (and whose disadvantaged) by this.

Glossary

Cross-references to other glossary terms are in **bold** typeface.

Accuracy in news reporting. A key concern in journalistic professional **ideology** and in objectivist approaches to the sociology of news, sitting alongside concerns about **bias**, **balance** and **objectivity** in news. The focus on accuracy has been particularly pronounced in studies of science news coverage. For a review of accuracy research, see Hansen (2016). The notion of accuracy assumes the existence of a 'correct' master account or narrative – in relation to science/environment coverage, the correct version is often assumed to be the scientist's account or the scientific paper on which a news report is based. The notion of accuracy has little or no meaning in **constructionist** approaches to news, as the interest here is primarily in mapping the dynamics of various contending accounts/explanations/discourses and the way in which such accounts are promoted, elaborated and cemented in public debate.

Advertising. The promotion or bringing to public notice of goods, information, images through – normally paid for – displays in any medium (e.g. the press, broadcast and other electronic media, billboards, cinema, etc.). Of particular interest in this book are: 1) the use of advertising by government, industry/corporations, NGOs and pressure groups to promote particular definitions/claims regarding environmental matters, and 2) the use of **nature** and natural imagery in the advertising and promotion of a wide variety of consumer goods.

Agenda-setting. In communication research, this term refers to the **power** of the news media to influence public perception of the relative prominence and importance of different events, issues and actors/agencies. Originally formulated as the **power** of the media to influence public perception of the hierarchy of issues – what the public thinks about as opposed to the specific 'for-or-against' direction of

public opinion – the term has increasingly overlapped with notions of **framing** to indicate the process by which the boundaries and hierarchy of public discourse are formed and shaped. The classic agenda-setting reference in communication research is McCombs and Shaw, 1972.

Alienation

> William Kornhauser in the *Politics of Mass Society* (US: Free Press, 1959) argues that the breakdown and decline of community groups and the extended family in modern society produces feelings of isolation and increases the possibility that people will be influenced by the appeals of extremist political groups. Alienation might therefore be a significant variable in determining an individual's receptivity to certain types of communication.
>
> (Watson & Hill, 2015: 7)

Alternative media. This can generally be considered as any medium which is not controlled or owned by business corporations or government. The Royal Commission on the Press (1977) delineates the following useful characteristics: alternative media deal with the opinions of minorities; express attitudes 'hostile to widely held beliefs' and give coverage to subjects and views which do not feature regularly in mainstream media. The term is useful for understanding the history and context of environmental activism in the 20th century but has less relevance in the changed digital media and communications landscape of the Internet and online communication in the 21st century.

Attention/legitimacy/action. Solesbury (1976) proposed these as the three key tasks for pressure group campaigning: 1) commanding attention (e.g. in the media); 2) claiming legitimacy (i.e. ensuring that the particular definitions and stance promoted are received and presented in public and/or media debate as legitimate and appropriate within the generally accepted terms of public debate, as opposed to being undermined or rejected as for example 'extremist'); and 3) invoking action in the form of political, policy, or legislative changes, or indeed in the form of **public opinion** and behaviour change. The particular usefulness of Solesbury's task-list is that it directs critical attention to the need for all three tasks to be met, that is, that getting media coverage alone is of little use to a pressure group if it's the 'wrong' kind of coverage or has no further implications in the form of political/policy change.

Balance/bias/accuracy in reporting. See **accuracy in news reporting** above.

Blue-chip programmes. A **genre** label used to denote high-cost and expensive-looking nature/wildlife television programmes, often contrasted with 'adventure/presenter-led' programmes. While acknowledging that definitions vary, Bousé (1998: 134) proposes the following core characteristics: the depiction of *mega-fauna*; *visual splendour*; *dramatic narrative*; and the *absence of history, politics, people* and *science*.

Carrying capacity. Hilgartner and Bosk (1988) point out that 'all public arenas, operatives, and members of the public have finite resources to allocate to social problems' (p.60). **Public arenas** such as the press and broadcast media, political institutions, institutions of government, the courts, public institutions, and so on, can only entertain or carry a limited number of issues on their agenda at any one point in time. The introduction of new issues on the agenda therefore inevitably has the effect of either reducing the space/time given to other issues already on the agenda or pushing them off the agenda altogether. Different arenas can cope with different numbers of issues at any one point in time. The way in which these finite capacities impact on the dynamics between different issues or **social problems** is particularly interesting: do environmental issues, for example, get squeezed off the public agenda during periods when the economy is in trouble?

Circulation of claims. The term is used here principally in contradistinction to traditional linear models of communication (sender > message/medium > audience/recipient), to indicate the interactive and dynamic nature of public debate. A drawback of the 'circulation' metaphor in this context is that it does not adequately express how claims and messages are *changed* as they circulate.

Claims-makers/issue sponsors. Any individual, group, agency or institution involved in making claims about or promoting/sponsoring issues, problem definitions or debate in **public arenas**, such as the media. The term claims-maker originates with Kitsuse and Spector's (1973: 415) definition of social problems as 'the activities of individuals or groups making assertions of grievances and claims with respect to some putative conditions'. In media and communications research, the term claims-maker is often used synonymously with **sources** and actors in news media content. Claims-makers are also frequently referred to as 'issue sponsors' and occasionally as 'issue entrepreneurs'.

Claims-making styles. One of Ibarra and Kitsuse's (1993) four rhetorical dimensions, claims-making styles concern the 'bearing and tone' with which claims are fashioned and presented. Examples include *scientific, comic, theatrical, civic, legalistic and subcultural styles.* While there is overlap with concepts such as discourse, **frame** and **genre**, the focus on style can be useful for understanding why some claims fare much better than others in public, and why some gain popularity and legitimacy more easily than others. (See also: **rhetorical idioms**, **counter-rhetorics**, **motifs** and **settings**.)

Claims-making tasks. See **Attention/legitimacy/action** above: 1) commanding attention; 2) claiming legitimacy; 3) invoking action.

Commodification. 'The transformation of relationships, formerly untainted by commerce, into commercial relationships, relationships of exchange, of buying and selling' (Encyclopedia of Marxism, 2009). Used here in relation to the use of **nature** symbolism and associated values for selling material products or goods for a profit. Originating in Marxist theory, the term refers to the assignment of economic or profit value to non-material goods such as concepts, ideas, identities or information. Here, the term is used to indicate the process by which advertisers translate the values and identities associated with **nature** or the countryside into purchaseable material goods.

Constructionist approach/constructionism. Originally formulated in relation to the analysis of social problems, the constructionist approach 'breaks with conventional and commonsensical conceptions of **social problems** by analyzing them as a *social process* of definition' (Miller & Holstein, 1993: 6). Spector and Kitsuse (1977/1987: 75–76) define **social problems** as '*the activities of individuals or groups making assertions of grievances and claims with respect to some putative conditions*', but the most pertinent part of their definition for media and communication research is where they go on to state that the '*central problem for a theory of social problems is to account for the emergence, nature, and maintenance of claims-making and responding activities*'. A constructionist approach to media and communication therefore focuses on the way in which issues, problems, claims and definitions emerge through social processes of communication, enter into and are elaborated in **public arenas** (notably the mass media), provoke and are met with **counterclaims/counter-rhetorics**, and so on. The main focus is on accounting for the process of claims-making, not to establish whether claims – or their representation and inflection in media and other **public arenas** – are **accurate, objective** or **balanced**.

Content analysis. A systematic and quantitative method for analysing media content, it involves the transparent and systematic coding and counting of specified dimensions or characteristics of content in selected samples of media output. Content analysis is one of the most widely used methods in media and communication research, and has been and continues to be prominently used in analyses of media coverage of environmental issues. Studies examining the longitudinal trends in media coverage of the environment have used content analysis extensively, often – and productively – in combination with qualitative approaches to media content (e.g. discourse analysis) and in combination with **survey** studies of **public opinion**.

Corporate image strategies. Corporate communication strategies designed to improve and promote a positive image of a corporation/ business in the public sphere and/or designed to engage with, counter or undermine campaigns or policies perceived as being against or restrictive to the interests of a corporation or business.

Counter-rhetoric/counterclaim. Several scholars have noted the simple **dialectics** of claims-making: for every claim, there is a counterclaim. Counter-rhetorics (Ibarra & Kitsuse, 1993) or counterclaims are discursive strategies for countering or undermining existing claims promoted by adversaries. As such, they are less concerned with the construction of thematically coherent claims, and instead focus on ways of chipping away at or undermining the credibility of existing claims. They are not simple 'opposites' in the sense that a claim is countered by claiming the direct opposite, but tend instead to focus on undermining **credibility**, exposing inconsistencies, questioning the robustness of evidence, ridiculing **claims-makers** or questioning their sincerity and credentials, and casting doubt on the validity of claims.

Credibility/trust in news reporting. While journalists, particularly environmental and science correspondents, who tend to stay with their specialist field longer than other types of reporters, may well build up considerable expertise and knowledge relating to their field of reporting, they are not 'authorised' or 'accredited' experts per se. The credibility of their reporting thus has to be actively constructed in news reports, and this is often done through quoting recognisable or recognised expert **sources** (people, institutions or published data). Journalists deploy a range of 'markers' for determining and conveying the standing and credibility of their **sources.** Fundamentally, the journalistic task of determining, assessing and conveying credibility is made considerably easier through the cultivation of trustworthy and trusted **sources** over time, that is, where evidence and information is supplied by a trusted **source**, there is little or no need to spend time checking the credentials of sources or the validity of information supplied.

Cultivation analysis. A prominent and influential form of media effects research originally introduced by American communications scholar George Gerbner. Its central proposition is that the long-term repetition of core message patterns in popular media results in a matching world view in media audiences. The more time one spends consuming media news and entertainment, the more one's beliefs and ideas will match those that dominate on television. In the environmental communications field, James Shanahan and colleagues have, since the early 1990s, been at the forefront of testing the relationship between media representations and audience beliefs and opinion regarding the environment and environmental issues.

Cultural packages/media packages. Cultural packages/media packages/**interpretive packages** refer to the notion that claims, arguments, opinions do not exist as mere compilations of atomised words and images, but rather as organised, structured clusters or packages. As Gamson and Modigliani suggest (1989: 3):

> A package has an internal structure. At its core is a central organizing idea, or frame, for making sense of relevant events, suggesting what is at issue. [...] This frame typically implies a range of positions, rather than any single one, allowing for a degree of controversy among those who share a common frame. Finally, a package offers a number of different condensing symbols that suggest the core frame and positions in shorthand, making it possible to display the package as a whole with a deft metaphor, catchphrase, or other symbolic device.

Cultural proximity (as a news value). One of several **news values** described by Galtung and Ruge (1965) in the now classic reference on the structure of news. Geographical, political and cultural proximity thus significantly enhances the likelihood of news coverage: people, events and issues in neighbouring countries of a similar culture are more likely to receive news coverage than their equivalents in countries or regions which are geographically, politically or culturally more distant.

Cultural resonance. The extent to which claims and news accounts are in harmony with, positioned within, activate or 'speak to' generally accepted cultural themes or **narratives**.

> Cultural resonances are used to shape generally recognisable plots (rags to riches, power corrupts). They offer easily recognized social/cultural stereotypes of characters (evil villains, honourable victims, noble heroes and heroines), and they reinforce general social goals, i.e., the underlying or implicit values that shape the way the mainstream media organize their impressions of society.
>
> (Ryan, 1991: 79)

Desk journalism (also desk-bound journalism). Journalistic work conducted from the journalist's desk (literally or metaphorically) and relying increasingly or principally on information gathering via digital communication, for example mobile phone or the Internet. This contrasts with the traditional image of the journalist as 'out and about' attending events and press conferences, and tracking down **sources** to be interviewed face-to-face.

Dialectic. Used here in the sense of spiralling dynamic interaction between claims and counterclaims and drawing directly on German philosopher Friedrich Hegel's (1770–1831) model of reasoning: a *thesis* is met with an *anti-thesis*, resulting in a new *synthesis* at a higher level, from which the process then repeats.

Direct media effects. A term used in media and communications research to indicate the media's direct influence on audience beliefs, perception, opinion and/or behaviour. The 'direct effects' model of media influence is associated with the early part of the 20th century and with **mass society** theory, but was severely questioned and criticised as early as the 1940s and 1950s, and is now generally discredited for its overly simplistic and linear view of media influence.

Disaster news. Predominantly used to refer to news about major accidents (chemical spills, oil spills, air crashes, nuclear power plant accidents, etc) or natural disasters – earthquakes, tsunamis, volcanic eruptions, floods, droughts, and so on – and their consequences for people, **nature**/wildlife and societies.

Echo chamber effect. 'Web publics are subject to fragmentation, insulation from broader sources of information and comment, in danger of being a number of *enclaves*, talking between themselves and only to themselves, in a chamber that only echoes their own discourses' (Watson & Hill, 2015: 88). A feature of the digital communications landscape that poses significant challenges for the public sphere ideal of publics making up their minds about, for example, environmental policy on the basis of exposure to and engagement with diverse arguments and evidence.

Event orientation. The tendency for news to be focused on events rather than issues, and on products/outcomes rather than processes. In media and communications research, the key criticism of the observed event orientation of news concerns the way in which the preoccupation with events and actions obscures or ignores the political and historical contextual information necessary for understanding and interpreting the meaning and implications of events/actions.

Excellent public relations function. A term used by Grunig et al., (2002) in their influential introduction to 'excellence' in public relations: 'An excellent public relations function integrates all public relations programs into a single department or provides a mechanism for coordinating programs managed by different departments' (Grunig et al., 2002: 15).

Forum. The physical **setting**, context or **public arena** (e.g. political institution, the courts, research establishment, the media) for claims-making or debate, or as a focus for media attention. The forum is not simply a neutral or inert stage, but – through its own format requirements and conventions – restricts and sets boundaries for what can be said and how.

Frame(s). Gitlin (1980) defines frames as 'principles of selection, emphasis, and presentation composed of little tacit theories about what exists, what happens, and what matters'. Frames draw attention to particular dimensions or perspectives, and in doing so they also set the boundaries for how what is presented is discussed, interpreted or perceived. 'News frames are almost entirely implicit and taken for granted. (…) News frames make the world look natural. They determine what is selected, what is excluded, what is emphasised. In short, news presents a packaged world' (Gamson, 1985: 618).

Framing

> To frame is to select some aspects of a perceived reality and make them more salient in a communicating text, in such a way as to promote a particular problem definition, causal interpretation, moral evaluation, and/or treatment recommendation for the item described.
>
> (Entman, 1993: 56)

Framing in media coverage involves *selection/accessing* of **sources/claims-makers** and *emphasis* in the presentation/evaluation of arguments/actors. Framing analysis can usefully proceed (see Ryan, 1991) by asking: 1) What is the issue? 2) Who is responsible? and 3) What is the solution?

Front groups. A group, organisation or coalition of groups created to mask the vested interests of companies, corporations, parties or general stakeholders. Front groups project in the public sphere the appearance of working objectively and for 'the common good'. A prominent example is the Global Climate Coalition, set up with the help of **public relations** experts by major oil-producing interests opposed to policies designed to curb CO2 emissions, by casting doubt on climate change as anthropogenic.

Gallup. Short for Gallup Poll: the

> trademark name for an assessment of public opinion by the questioning of a representative sample, typically as a basis for forecasting votes in an election. It is named after George H. *Gallup* (1901–84), the American statistician who devised the method.
>
> (Knowles, 2006)

Genre. A particular type or category of media content, characterised by recognisable conventions of style, form and presentation. Key media genres include: news, documentary, current affairs, chat show, reality show, **advertising**, drama serial, comedy, editorial, letter to the editor, feature/opinion article, and so on. Genre categories are fluid and flexible rather than absolute, and are often deliberately combined (e.g. 'docudrama') or manipulated for effect.

Globalisation

> Globalisation theory examines the emergence of a global cultural system. It suggests that global culture is brought about by a variety of social and cultural developments: the existence of a world satellite information system; the emergence of global patterns of consumption and consumerism; the cultivation of cosmopolitan life-styles; the emergence of global sport such as the Olympic Games, world football competitions, and international tennis matches; the spread of world tourism; the decline of the sovereignty of the nation-state; the growth of a global military system; recognition of a world-wide ecological crisis; the development of world-wide health problems such as AIDS; the emergence of world political systems such as the League of Nations and the United Nations; the creation of global political movements such as Marxism; extension of the concept of human rights; and the complex interchange between world religions. More importantly, globalism involves a new consciousness of the world as a single place.
>
> (Scott & Marshall, 2009)

Greenwashing

> The unjustified appropriation of environmental virtue by a company, an industry, a government, a politician or even a non-government organization to create a pro-environmental image, sell a product or a policy, or to try and rehabilitate their standing with the public and decision makers after being embroiled in controversy.
>
> (Sourcewatch, 2017)

Hegemony/hegemonic

> A term introduced by the early-twentieth-century Italian Marxist theorist Antonio Gramsci to describe a certain kind of **power** that arises from the all-embracing ideological tendencies of mass media to support the established **power** system and exclude opposition and competing values. In brief it is a kind of dominant consensus that works in a concealed way without direct coercion.
>
> (McQuail, 2010: 558)

Hegemony: 'leadership or dominance, especially by one state or social group over others.' Hegemonic: 'ruling or dominant in a political or social context: *the bourgeoisie constituted the hegemonic class*' (Soanes & Stevenson, 2008).

Hyperlinking. Linking from a website or document to another website or file. The presence of hyperlinks on a **claims-maker**'s website should be seen as a deliberate part of claims-making strategy and, at the very least, presents an acknowledgement of the presence of the (possibly opposing or alternative) positions of claims expressed at the sites linked to.

Identity. 'Specific characterisation of person, place, and so on by self and others, according to biographical, social, cultural and other features. Communication is a necessary condition for forming and maintaining identity. By the same token, it can weaken or undermine it.' (McQuail, 2010: 559)

Ideology. 'A cohesive set of beliefs, ideas, and symbols through which persons interpret the world and their place within it' (Calhoun, 2002). The underlying world view, value system or perspective which informs, and, to some extent, governs the nature and surface manifestation of communication. Ideology, like framing, works effectively by 'naturalising' the view or values that it expresses. The interests served by particular ideologies – that is, who (individuals, institutions, groups, social classes, etc.) benefits from this particular way of 'looking at' or 'defining' the issues, events or actors communicated about – are communicated principally through: (a) the particular choice of words/metaphors used in a communicating text; and (b) the way in which a text is structured both horizontally, i.e. the **narrative**, and vertically, i.e. the juxtaposition of characters, events and issues within the text.

Image enhancement/image management. The strategic use of **advertising**, **public relations** and other planned communication to improve the public image or identity of a corporation, group, political party or public agency.

Information subsidy. The act or process of **sources** providing to journalists or news organisations ready-packaged information that can be easily adapted to the format and other requirements of news. Originating in Oscar Gandy's (1982) influential study *Beyond Agenda Setting: Information Subsidies and Public Policy*, the term is widely used in studies of the relationship between **sources** and journalists in environmental, science and health reporting. Press releases and PR are among the most obvious forms of information subsidy.

Insider/outsider groups. 'Insider groups are regarded as legitimate by government and are consulted on a regular basis. Outsider groups either do not wish to become enmeshed in a consultative relationship with officials, or are unable to gain recognition' (Grant, 2000: 19). The distinction is helpful for appreciating first and foremost that not all pressure groups or NGOs (non-governmental groups) are keen on publicity or see the media as a primary campaigning **forum**. Second, the distinction is helpful for sharpening our understanding of the significance and use of news media to the communication and campaigning strategies of outsider groups, particularly in terms of garnering both financial and political support for their cause.

Intelligence gathering and surveillance as pressure group strategies. While environmental pressure groups often come to public attention through spectacular and newsworthy media stunts and performances, the real key to success is in the meticulous and resource-demanding (i.e. often beyond the resources available to news organisations or individual journalists) surveillance and gathering of intelligence regarding developments in environmental policy **forums**, political negotiations, legislation and so on, followed by carefully targeted dissemination (e.g. to news organisations, through **information subsidies**) of appropriately framed information.

Interpretive packages

> Media discourse can be conceived of as a set of interpretive packages that give meaning to an issue. A package has an internal structure. At its core is a central organizing idea, or frame, for making sense of relevant events, suggesting what is at issue. [...] a package offers a number of different condensing symbols that suggest the core frame and positions in shorthand, making it possible to display the package as a whole with a deft metaphor, catchphrase, or other symbolic devices.
>
> (Gamson & Modigliani, 1989: 3)

IPCC: Intergovernmental Panel on Climate Change. The international body for assessing the science related to climate change. The IPCC was set up in 1988 by the World Meteorological Organization (WMO) and United Nations Environment Programme (UNEP) to provide policymakers with regular assessments of the scientific basis of climate change, its impacts and future risks, and options for adaptation and mitigation.(IPCC, 2018)

Issue packages. See **interpretive packages** above.

Issue-attention cycle. Downs (1972) proposed the label 'issue-attention cycle' to describe the cyclical manner in which various **social problems** suddenly emerge on the public stage, remain there for a time, and 'then – though still largely unresolved – gradually [fade] from the centre of public attention' (p.38). Downs identified five distinctive stages in the issue-attention cycle: (1) a pre-problem stage; (2) alarmed discovery and euphoric enthusiasm; (3) realising the cost of significant progress and the sacrifices required to solve the problem; (4) gradual decline of intense public interest; and (5) the post-problem stage, where the issue has been replaced at the centre of public concern and 'moves into a prolonged limbo – a twilight realm of lesser attention or spasmodic recurrences of interest.'

Mass society

> A description of modern, industrial society as a mass of undifferentiated and alienated individuals. Mass society became an object of concern in the early nineteenth century and initially reflected a shift in the nature of elitist fears for the body politic. Where the 'tyranny of the majority' once expressed fears of disruptive mobs and demagogic rule, the new forces of modernisation implied stronger leveling tendencies that threatened to eliminate the values traditionally identified with social aristocracy – especially excellence and individuality. Fear of the mob gave way to fear of the conformist, degraded mass. [...] Marxist and renewed liberal versions of mass-society critique emerged largely in response to Europe's authoritarian turn in the 1930s. In an effort to explain the appeal of Nazism, fascism, and communism, liberal social scientists such as David Riesman (*The Lonely Crowd*, 1950) and William Kornhauser (*The Politics of Mass Society*, 1960) emphasized the decline of traditional religious and moral attachments, and the rise of sophisticated propaganda techniques that could manipulate the mass and achieve consent. The Marxist Frankfurt school contended as early as the 1940s that a mass society of alienated individuals was the inevitable product of a culture industry that served the interests of capitalism.
>
> (Calhoun, 2002)

Motifs. One of Ibarra and Kitsuse's (1993: 47) five foci for analysis of claims-making, motifs are defined as recurrent thematic elements, metaphors and figures of speech that encapsulate, highlight or offer a shorthand to some aspect of a **social problem**. Examples: epidemic, menace, scourge, crisis, blight, casualties, tip of the iceberg, the war on (drugs, poverty, crime, gangs, etc.), abuse, hidden costs, scandal, ticking time bomb.

Multimodal analysis. In communication research, an analytical approach that recognises that communication increasingly comprises not just textual/verbal and sound/aural modes but particularly visual modes, and that these modes must be analysed together to establish the meaning and implications of communication. Multimodal analysis in communication research draws primarily on the analytical tools and insights of linguistics and semiotics.

Myth

> Generally refers to stories that contribute to the elaboration of a cosmological system and to a cohesive social identity – e.g., accounts of origins, explanations of values and taboos, and narrative legitimations of authority. […] Claude Lévi-Strauss is largely responsible for the structuralist approach to myth as a network of interchangeable narrative elements (mythemes) that reveal the basic oppositions that organize a given culture (endogamy and exogamy, animal and vegetable, raw and cooked, and so on). Roland Barthes gave the term a different and highly influential inflection in his book *Mythologies* (1957). For Barthes, myths are the codes that underlie the imagery and practices of much of contemporary culture. Their primary function is to lend the appearance of universality to otherwise contingent cultural beliefs. In this respect, myth occupies much the same place in Barthes's work as ideology in the writing of Antonio Gramsci and Louis Althusser: it naturalizes and secures consent for the status quo.
>
> (Calhoun, 2002)

Narrative

> A telling of some true or fictitious event or connected sequence of events, recounted by a narrator to a narratee (although there may be more than one of each).[…] A narrative will consist of a set of events (the story) recounted in a process of narration (or discourse), in which the events are selected and arranged in a particular order (the plot).
>
> (Baldick, 2008)

Narrative analysis is useful in media and communication analysis perhaps particularly because it strips away distinctions between genres (news, documentary, advertising, entertainment, etc.) and focuses on the sequencing of events and the positioning/role/relationships of the actors who drive the narrative/story forward.

Natural history models of issue careers. An analytical frame implying that **social problems** develop or pass through a set of sequentially ordered stages.

> Social problems do not arise full-blown, commanding community attention and evoking adequate policies and machinery for their solution. On the contrary, we believe that social problems exhibit a temporal course of development in which different phases or stages may be distinguished. Each stage anticipates its successor in time and each succeeding stage contains new elements which mark it off from its predecessor. A social problem […] passes through the natural history stages of awareness, policy determination, and reform.
>
> (Fuller & Myers, 1941: 321)

Nature documentary/nature programmes. The term 'nature documentary' is widely and broadly used to refer to any (mainly television) non-fiction, informative programme about some aspect of nature. It is distinguished by (generally) adhering to the format and **genre** conventions of 'documentary' and by its content focus on wildlife, plants and other aspects of the natural environment (as opposed to documentaries about social, political or cultural issues). However, as Bousé (1998) indicates by his rhetorically titled article 'Are wildlife films really "nature documentaries"?' there is considerable room for further distinctions within the broad label 'nature documentary', particularly with regard to the differences in storytelling conventions/narrative formats deployed in **wildlife films** and in nature documentaries (see also **blue-chip programmes** and **wildlife films**).

Nature/natural. While both 'nature' and 'natural' in common usage inherently suggest an ontological quality, their meaning is of course socially, historically and culturally constructed. Williams (1983: 219) distinguishes three principal meanings of 'nature': 'i) the essential quality and character of something; ii) the inherent force which directs either the world or human beings or both; (iii) the material world itself, taken as including or not including human beings.'

News cycle. The length of time between each edition of a news outlet: for daily (printed) newspapers, 24 hours; for traditional radio and television channels, the number of hours between each major news programme. The significance of the news cycle has traditionally related to questions about immediacy, **agenda-setting** and competition between news media/organisations. With the advent of 24-hour news channels and online news, news is potentially updated on a 'rolling' and continuous basis, making the notion of a 'cycle' less immediately relevant, although a certain 'rhythm' can still be apparent. Awareness of 'news cycles' is important to **sources/claims-makers** in relation to timing and targeting claims for optimum effect/impact.

News forum/setting/arena. In principle, any (newsworthy) **setting**, context or **public arena** (e.g. political institution, the courts, research establishment, the media) for claims-making or debate. In practice, the term tends to refer to the particular standard settings or arenas that journalists routinely attend to or monitor for news stories (particularly the institutions of government, international organisations, the courts, research and information/knowledge-producing establishments, etc.).

News values. The set of criteria that journalists and news media use for determining whether to report an event or story. News values vary according to cultural context and target audience. The standard reference is Galtung and Ruge's (1965) classic study 'The Structure of Foreign News', which lists a dozen or so criteria, including *frequency, negativity, unambiguity, meaningfulness* (***cultural proximity*** and *relevance*), *continuity, unexpectedness,* and so on.

Newsworthiness. The degree to which events/stories meet the **news values** criteria. Frequently used synonymously with **news values**.

Nostalgia. 'A sentimental longing or wistful affection for a period in the past: [...] Something done or presented in order to evoke such feelings: *an evening of TV nostalgia.* [...] from Greek *nostos* "return home" + *algos* "pain"' (Soanes & Stevenson, 2005).

> Nostalgia became, in short, the means for holding onto and reaffirming identities which had been badly bruised by the turmoil of the times. In the "collective search for identity" which is the hallmark of this postindustrial epoch – a search that in its constant soul-churning extrudes a thousand different fashions, ecstasies, salvations, and utopias – nostalgia looks backward rather than forward, for the familiar rather than the novel, for certainty rather than discovery.
>
> (Davis, 1979: 107–108)

Objectivity. A journalistic professional value that goes together with, and is frequently seen as synonymous with, the professional journalistic news requirements of **accuracy**, fairness, transparency, impartiality, separation of fact from comment, **balance** and lack of **bias**. Critics argue that objectivity in news reporting is impossible and that the journalistic construction of the appearance of objectivity is itself a concealment, whether intended or not, of **bias**.

Paradigmatic and syntagmatic analysis. Syntagmatic analysis (also referred to as narrative analysis) is the study of the linear structure of storytelling. It is concerned with how meaning arises from the sequential arrangement of words or actions in a text, where a change in sequence may/will result in a different meaning. By contrast, paradigmatic analysis (also referred to as structural analysis) studies the relationship between the words that appear in a text and the reservoir of other words, not chosen, in the underlying language, and it studies the way that meaning is communicated through the structural arrangement and juxtapositions of actors, values and events in a text.

Postmodernism. A 'late 20th-century style and concept in the arts, architecture, and criticism, which represents a departure from modernism and is characterized by the self-conscious use of earlier styles and conventions, a mixing of different artistic styles and media, and a general distrust of theories' (Soanes & Stevenson, 2005). In Berger's

(2016: 395) words, 'the old philosophical belief systems that had helped people order their lives and societies are no longer accepted or given credulity. This has lead to a period in which, more or less, anything goes.' This characterisation has particular pertinence for environmental, risk and similar science-based communication, where earlier trust and belief in scientific and political authority has eroded. In journalism, news and documentary programming, the postmodernist trend manifests itself as an erosion of traditional journalistic values of factual **accuracy**, impartiality and **objectivity**, but more significantly in a move from 'visual realism' towards a mixing of styles and referencing of other media/texts rather than a 'window-on-the-world' referencing of the 'reality' being portrayed.

Power. A complex concept with many meanings across the various disciplines of the social sciences, but of key interest in the present book is its meaning as the possession of necessary resources (economic, technical or indeed communicative/cultural competence or capital) and associated ability to effect or bring about change – or to prevent change from happening – by influencing and manipulating **claims-making**, media and communications agendas, including the ability to prevent certain claims from making it onto the agendas of the media and other **public arenas**.

Primary definers. The **claims-makers** or **sources** who influence and shape, through direct quotation or indirect referencing, the media and news agenda. The journalists and media themselves are often referred to as secondary definers. The term primary definer can be and has been used to refer to any source quoted or referred to in media and news content, but in its original definition it was implied or assumed that primary definers were in a position of authority and **power** in society.

Progress package. The progress package (Gamson & Modigliani, 1989) is characterised by beliefs in science, technological innovation, mastery over **nature**, efficiency, economic expansion, adaptability, practicality, expediency, and so on, as the solution to problems and the route to a better, safer and more prosperous society. One of its clearest manifestations in public debate about climate change is in the form of proposals to 'manage' and control climate change through technological innovation, that is, to 'invent' ourselves out of trouble.

Public arenas model. A model introduced by Hilgartner and Bosk (1988) for the analysis and understanding of the processes and **forums** of claims-making and social problems construction:

> The collective definition of social problems occurs not in some vague location such as society or **public opinion** but in particular public arenas in which social problems are framed and grow. These arenas include the

> executive and legislative branches of government, the courts, made-for-TV movies, the cinema, the news media (television news, magazines, newspapers, and radio), political campaign organisations, social action groups, direct mail solicitations, books dealing with social issues, the research community, religious organisations, professional societies, and private foundations. It is in these institutions that social problems are discussed, selected, defined, framed, dramatized, packaged, and presented to the public.
>
> (58–59)

The model directs attention to the commonalities (e.g. limited **carrying capacities**) and differences (e.g. timetable/time cycles) across the major public arenas. It points to the centrality of factors such as competition, selection, format, the routines of operatives, and so on, in influencing claims-making processes and it highlights the ways in which effective **claims-makers** tailor and adapt their claims to fit the requirements of public arenas.

Public opinion

> An ill-defined concept, used in many ways, but perhaps most generally it refers to the approval or disapproval of publicly observable positions and behaviour, as expressed by a defined section of a society, and (usually) measured through opinion polls. Consequently, it is often taken to be synonymous with 'what the polls report' – about morality, favoured consumer brands, politics, or whatever.
>
> (Scott & Marshall, 2009)

Public relations/PR

> Now a reference to all forms of influence carried out by professional paid communicators on behalf of some 'client' and designed primarily to project a favourable image and to counter negative views that might exist. The means are various, ranging from direct communication to providing gifts and hospitality. Public relations is often a source of supply for news media or seeks to influence news in other ways.
>
> (McQuail, 2010: 568)

Public sphere

> The conceptual 'space' that exists in a society outside the immediate circle of private life and the walls of enclosed institutions and organisations pursuing their own (albeit sometimes public) goals. In this space, the possibility exists for public association and debate leading to the formation of **public opinion** and political movements and parties that can hold private interests accountable. The media are now probably the key institution of the public sphere, and its 'quality' will depend on the quality of media. Taken to extremes, certain structural tendencies of media, including concentration, commercialisation and globalisation, are harmful to the public sphere.
>
> (McQuail, 2010: 569)

Publicity stunts. A demonstration or performance enacted for the purpose of drawing media and public attention, and for the purpose of getting news media coverage. Publicity stunts are specifically designed to appeal to and exploit core **news values** (unexpectedness, visuality, drama, etc.). On their own, they are of limited use to pressure groups, but as part of a wider campaign strategy they can be a highly effective way of drawing attention to campaign issues.

Realism/realist. In both factual and fictional media content, a mode of representation which is or appears to be accurate, objective, true to life or a 'window on the world'. Conventional news journalism is 'realist' in the sense that it reports the 'objective' facts in an impartial way, as an impartial observer of real events, people and issues without drawing attention to its own constructedness and **genre** conventions. **Nature documentaries** which purport to be simply filming and observing wildlife and natural events 'as they occur' can be described as 'realist' or as adhering to the conventions of realism. Television drama serials or 'soap operas' purporting to portray 'real' people, living in broadly recognisable 'real' environments and dealing with real issues can likewise be characterised as adhering to a realist mode of presentation. The core notion of television merely 'observing life as it happens' is also evident in the **genre**-label reality show/reality television.

Rhetoric

> The art of effective or persuasive speaking or writing, especially the exploitation of figures of speech and other compositional techniques. Language designed to have a persuasive or impressive effect, but which is often regarded as lacking in sincerity or meaningful content: *all we have from the Opposition is empty rhetoric.*
>
> (Soanes & Stevenson, 2005)

Rhetorical idioms. Ibarra and Kitsuse (1993: 34) advocate that **constructionist** studies should distinguish 'four overlapping but analytically distinct rhetorical dimensions: rhetorical idioms, **counter-rhetorics**, **motifs**, and **claims-making styles**.' While a dictionary definition of an idiom (literally: peculiar phraseology) is 'a group of words established by usage as having a meaning not deducible from those of the individual words (e.g. *over the moon*, *see the light*)' or 'a form of expression natural to a language, person, or group of people' (Soanes & Stevenson, 2005), Ibarra and Kitsuse's use of the term 'rhetorical idioms' is wider and refers to the general cluster of words and rhetoric that characterises a particular perspective, moral evaluation or way of 'talking about' a social issue/problem. In this respect, their notion of rhetorical idioms is akin to **cultural packages**.

> *Rhetorical idioms* are definitional complexes, utilizing language that situates condition-categories in moral universes.[...] Each rhetorical idiom calls forth or draws upon a cluster of images. The *rhetoric of loss,* for example, evokes symbols of purity and tends toward nostalgic tonalities. The *rhetoric of unreason* evokes images of manipulation and conspiracy. The *rhetoric of calamity* situates condition-categories amid narratives of widespread devastation, and so on.
>
> (Ibarra & Kitsuse, 1993: 34)

Romanticism. 'A movement in the arts and literature which originated in the late 18th century, emphasizing inspiration, subjectivity, and the primacy of the individual' (Soanes & Stevenson, 2008). Characterised by its emphasis on **nature** as pure, good, spiritual, sublime, authentic and pristine, the romantic view contrasted with the earlier Enlightenment period's view of **nature** as something wild and threatening to be studied, understood, tamed and controlled in the name of civilisation and progress. Both the romantic view and the **utilitarian** Enlightenment view continue to inform and influence the constructions of **nature** and the environment in media and public discourse.

Schema. 'Refers to the preconceived frame or script which is typically available to journalists for reporting isolated cases or events. A schema is an aid to communication and understanding, because it provides some wider context and sense-making' (McQuail, 2010: 569).

Science fiction

> A genre of fiction based on imagined future technological or scientific advances, major environmental or social changes, etc., and frequently portraying space or time travel and life on other planets. Science fiction emerged in the late 19th century in the works of writers such as Jules Verne and H.G. Wells, although there are earlier precedents, such as Mary Shelley's *Frankenstein* (1818).
>
> (Deverson, 2004)

Scripts. Akin to **cultural packages, narratives, schema** and **frames**, scripts can be regarded as the world views or clusters of meaning/perspective from our cultural reservoir, which help us make sense of our environment and understand how things work, what counts as appropriate or acceptable within our culture. Turney (1998), Huxford (2000) and others have persuasively argued that environment and science correspondents as well as journalists generally rely heavily on readily available cultural scripts and frames, particularly when reporting on new and unfamiliar developments in science and environmental issues.

Settings. Ibarra and Kitsuse (1993) add settings to their four rhetorical dimensions of claims-making (**rhetorical idioms**, **counter-rhetorics**, **motifs** and **claims-making styles**) as an important focus for analysing the construction of **social problems**. Settings – like **forums** and **arenas** – are the physical or abstract (e.g. academia) context in which claims-making is enacted or performed. Settings have important format, time and **genre** conventions, which structure what is said, how it is said, and indeed how that, which is said, is received, consumed or responded to by the public and by **claims-makers** in other forums. Ibarra and Kitsuse (1993: 54), pointing to the media as an important setting and touching on the format and **genre** conventions which govern different media **genres**, ask: 'What are the explicit or tacit rules for admissible testimony, fairness, objectivity, and so on? How does the visual component of some of these media alter the claim's sense, reception, and structure?' See also **News forum/setting/arena.**

Social constructionism. An approach in sociology and other disciplines which focuses on how our knowledge about the world around us is the result of social processes of definition. It directs attention to the analysis of processes of communication and definition, and to the analysis of the people, media and settings involved in articulating, elaborating, contesting and maintaining claims or definitions (see also **claims-making/claims-makers** and **social problems**). While media and communication processes are important foci in the wider social **constructionist** approach to the analysis of its main focus of concern, **social problems**, the **constructionist** framework has also become prominent as a framework and analytical approach in the sociology of news. Here, it rejects classic concerns about **bias**, **balance**, **accuracy** and **objectivity** in news journalism, and instead directs attention to the roles of **sources**, journalists and media organisations in the shaping of news content.

Social problems. Kitsuse and Spector (1973: 415) define social problems as 'the activities of individuals or groups making assertions of grievances and claims with respect to some putative conditions'. They reject (structural functionalist) sociological formulations which regard social problems as objective, identifiable conditions in society, and argue instead that social problems are the result of social processes of claims-making and definition. Their analytical emphasis is therefore on the rhetorical, discursive and definitional practices of **claims-makers**, on the rhetorical construction of claims (about social problems), and on the processes through which particular problem definitions are elaborated, contested and maintained in **public arenas**.

Source/news source

> An individual, group or institution that originates a message. In media terms, the source is where information starts, and it is an axiom of good reporting to ensure that the material supplied by the source is reliable and true. Best practice suggests that single sources be checked against other sources.
>
> (Watson & Hill, 2015: 301)

Strategic communication

> Examines how organizations use communication purposefully to fulfill their mission. All types of organizations, including private and public sector organizations, political parties, NGOs, and social movements, use strategic communication to reach their goals. Centrally, strategic communication examines the communicative practices of various types of organizations from an integrated perspective.
>
> (Frandsen & Johansen, 2017: 1)

Survey. In the general sense: an overview, examination or detailed description of someone or something. In the social sciences, the term mainly, although not exclusively, refers to the systematic collection of data/information, mainly with the use of a questionnaire, about a defined sample of a larger population, for example a survey of the environmental attitudes of teenagers.

Symbolic annihilation. A term

> used to highlight the erasure of peoples in popular communication. George Gerbner coined the term to describe the "absence" (1972, 44; Gerbner & Gross 1976 [...]), "condemnation," or "trivialisation" (Tuchman 1978, 17) of a particular group in the media. Generally applied to women and racial and sexual minorities, symbolic annihilation points to the ways in which poor media treatment can contribute to social disempowerment and in which symbolic absence in the media can erase groups and individuals from public consciousness. [...] Language use in the media also contributes to the trivialisation and condemnation of racial groups such as black people in popular communication.
>
> (Coleman & Yochim, 2008)

Time/cycle in news. See **News cycle**.

Uncertainty in news reporting. Much attention in studies of science, environment, health and risk communication has focused on the difficulties for journalists, media, politicians, scientists and experts in the handling and management in public communication of the fundamental degree of uncertainty and speculation, characteristic of most emerging scientific, environmental, health, social issues or problems.

Where in the past there may have been an unrealistic expectation that expert **sources** and the media could provide clear-cut answers and advice in relation, for example, to major accident, disease or public safety emergencies, there now seems to be a greater recognition – both in the media and in the public – of the complexities of these phenomena and of the difficulty of prediction.

Utilitarian. 'Designed to be useful or practical rather than attractive' (Soanes & Stevenson, 2005). A utilitarian (in contrast to **romantic** or conservationist) perspective on **nature** is one which sees nature and the environment as something to be mastered, controlled and exploited for the benefit of mankind. The utilitarian view of **nature** was prominent in the Enlightenment period of 18th-century Europe, and continues to be a prominent discourse throughout the 20th century, particularly in the immediate post World War II period, and to today.

Utopia/dystopia – and technopia. Utopia and dystopia are what semiologists call 'binary opposites' – one only makes sense/has meaning in relation to the other; when we talk of one, the presence of the other is always implicit or 'understood'. The *Oxford English Dictionary* (Soanes & Stevenson, 2005) defines utopia as 'an imagined place or state of things in which everything is perfect' and dystopia as 'an imagined place or state in which everything is unpleasant or bad, typically a totalitarian or environmentally degraded one.' Utopian and, perhaps more frequently, dystopian visions of the future are prominent implicit or explicit themes in literature, film, **advertising** and other media content. As such, they can often also be seen to inform or influence news reporting on environment, science and technology issues. Rutherford (2000), in his analysis of corporate **advertising**, introduces the further category *technopia* as 'the corporate version of a technological utopia' (190), corporate **advertising** projecting the idea that science and more particularly technology are keys to a bright, harmonious and prosperous future free of the environmental, social and other problems of the present.

Video news release (VNR). A video recording created by a pressure group, government department, **advertising** agency, PR firm, business, corporation or other **source/claims-maker** for distribution to news organisations. While originally referring to videotape recordings, they are now digital recordings. VNRs are the video equivalent of press releases and a visual type of **information subsidy**.

Visual environmental communication research. A growing field of research concerned with theorizing and empirically examining how visual imagery contributes to the increasingly multimodal public communication of the environment. Focused on a sociological understanding of the contribution that visuals make to the social, political, and cultural construction of "the environment," visual environmental communication research analytically requires a multimodal approach, which situates analysis of the semiotic, discursive, rhetorical, and narrative characteristics of visuals in relation to the communicative, cultural, and historical contexts and in relation to the three main sites—production, content, and audiences/consumption—of communication in the public sphere.'

(Hansen, 2017b)

Wildlife film. The portrayal or depiction of wildlife (and sometimes domestic animals) through the medium of film, defining characteristics include the narrative pretence of unmediated observation (a 'window' on wildlife going about its business) and (see Bousé, 1998: 134): the depiction of *mega-fauna*; *visual splendour*; *dramatic narrative*; and the *absence of history, politics*; *people* and *science*. Bousé argues for a distinction between wildlife films and **nature documentaries**: 'wildlife films are not documentaries; [...] they are primarily narrative entertainments that usually steer clear of real social and environmental issues' (p.xiv). See also **'blue-chip' programmes** and **nature documentary/nature programmes**.

References

Ader, C. R. (1995). A longitudinal study of agenda setting for the issue of environmental pollution. *Journalism & Mass Communication Quarterly, 72*(2), 300–311.

Advertising Standards Authority (ASA), & Committees of Advertising Practice (CAP). (2018). Advertising Codes. Retrieved 11 March, 2018, from https://www.asa.org.uk/codes-and-rulings/advertising-codes.html

Ahern, L., Bortree, D. S., & Smith, A. N. (2012). Key trends in environmental advertising across 30 years in National Geographic magazine. *Public Understanding of Science*. doi: 10.1177/0963662512444848

Albaek, E., Christiansen, P. M., & Togeby, L. (2003). Experts in the mass media: researchers as sources in Danish daily newspapers, 1961–2001. *Journalism & Mass Communication Quarterly, 80*(4), 937–948.

Aldridge, M., & Dingwall, R. (2003). Teleology on television? Implicit models of evolution in broadcast wildlife and nature programmes. *European Journal of Communication, 18*(4), 435–455.

Allan, S. (2006). *Online News: Journalism and the Internet*. London: Open University Press.

Allan, S., & Ewart, J. (2015). Citizen science/citizen journalism: new forms of environmental reporting. In A. Hansen & R. Cox (Eds.), *The Routledge Handbook of Environment and Communication* (pp. 186–196). London and New York: Routledge.

Anderson, A. (1997). *Media, Culture and the Environment*. London: UCL Press.

Anderson, A. (2014). *Media, Environment and the Network Society*. Basingstoke and New York: Palgrave Macmillan.

Anderson, A. (2015). News organisation(s) and the production of environmental news. In A. Hansen & R. Cox (Eds.), *The Routledge Handbook of Environment and Communication* (pp. 176–185). London and New York: Routledge.

Armitage, K. C. (2003). Commercial Indians: authenticity, nature and industrial capitalism in advertising at the turn of the twentieth century. *Michigan Historical Review, 29*(2), 71–94.

Atkinson, L. (2017). Portrayal and Impacts of Climate Change in Advertising and Consumer Campaigns. *Oxford Research Encyclopedia of Climate Science*. Retrieved 23 Aug. 2017, from http://climatescience.oxfordre.com/view/10.1093/acrefore/9780190228620.001.0001/acrefore-9780190228620-e-376.

Atkinson, L., & Kim, Y. (2015). "I drink it anyway and I know I shouldn't": understanding green consumers' positive evaluations of norm-violating non-green products and misleading green advertising. *Environmental Communication, 9*(1), 37–57. doi: 10.1080/17524032.2014.932817

Atwater, T., Salwen, M. B., & Anderson, R. B. (1985). Media agenda-setting with environmental issues. *Journalism Quarterly, 62*, 393–397.

Bagust, P. (2008). 'Screen natures': special effects and edutainment in 'new' hybrid wildlife documentary. *Continuum: Journal of Media & Cultural Studies, 22*(2), 213–226.

Bakir, V. (2006). Policy agenda setting and risk communication – Greenpeace, Shell, and issues of trust. *Harvard International Journal of Press/Politics, 11*(3), 67–88.

Baldick, C. (2008). *The Oxford Dictionary of Literary Terms*. Oxford: Oxford University Press: Oxford Reference Online.

Banerjee, S., Gulas, C. S., & Iyer, E. (1995). Shades of green: a multidimensional analysis of environmental advertising. *Journal of Advertising, 24*(2), 21–31.

Barthes, R. (1977). Introduction to the Structural Analysis of Narratives. In R. Barthes (Ed.), *Image, Music, Text: Essays Selected and Translated by Stephen Heath* (pp. 79–129). London: Fontana.

Bauer, M. (1998). The medicalization of science news – from the "rocket-scalpel" to the "gene-meteorite" complex. *Social Science Information Sur Les Sciences Sociales, 37*(4), 731–751.

Bauer, M., Durant, J., & Gaskell, G. (Eds.). (1999). *Biotechnology in the Public Sphere: A European Source-Book*. London: The Science Museum.

Bauer, M. W. (2002). Controversial medical and agri-food biotechnology: a cultivation analysis. *Public Understanding of Science, 11*(2), 93–111.

Bauer, M. W., Howard, S., Ramos, Y. J. R., Massarani, L., & Amorim, L. (2013). Global Science Journalism Report: Working Conditions & Practices, Professional Ethos and Future Expectations. London: SciDev.Net.

Baum, L. M. (2012). It's not easy being green… or is it? A content analysis of environmental claims in magazine advertisements from the United States and United Kingdom. *Environmental Communication–a Journal of Nature and Culture, 6*(4), 423–440. doi: 10.1080/17524032.2012.724022

Beck, U. (1992). *Risk Society: Towards a New Modernity* (M. Ritter, Trans.). London: Sage.

Beder, S. (1997). *Global Spin: The Corporate Assault on Environmentalism*. Totnes, Devon: Green Books.

Beder, S. (2002). *Global Spin: The Corporate Assault on Environmentalism* (Revised ed.). Totnes, Devon: Green Books.

Bennett, W. L. (1990). Toward a theory of press-state relations in the United States. *Journal of Communication 40(2) 103–125.*

Berger, A. A. (2016). *Media and Communication Research Methods: An Introduction to Qualitative and Quantitative Approaches* (4th revised ed.). London: Sage.

Besley, J. C., & Shanahan, J. (2004). Skepticism about media effects concerning the environment: examining Lomborg's hypotheses. *Society & Natural Resources, 17*(10), 861–880.

Best, J. (Ed.). (1995). *Images of Issues: Typifying Contemporary Social Problems* (2nd ed.). New York: Aldine de Gruyter.

Best, J. (2013). Constructionist social problems theory. In C. T. Salmon (Ed.), *Communication Yearbook 36* (Vol. 36, pp. 236–269). London: Routledge.

Blumer, H. (1971). Social problems as collective behavior. *Social Problems, 18*(3), 298–306.

Bortree, D. S., Ahern, L., Smith, A. N., & Dou, X. (2013). Framing environmental responsibility: 30 years of CSR messages in National Geographic Magazine. *Public Relations Review, 39*(5), 491–496. doi: 10.1016/j.pubrev.2013.07.003

Bousé, D. (1998). Are wildlife films really "nature documentaries"? *Critical Studies in Mass Communication, 15*(2), 116–140.

Bousé, D. (2000). *Wildlife Films.* Philadelphia: University of Pennsylvania Press.

Boyce, T. (2006). Journalism and expertise. *Journalism Studies, 8*(6).

Boyce, T., & Lewis, J. (Eds.). (2009). *Climate Change and the Media.* Oxford: Peter Lang.

Boykoff, M. (2011). *Who Speaks for the Climate?: Making Sense of Media Reporting on Climate Change.* Cambridge: Cambridge University Press.

Boykoff, M. T., & Boykoff, J. M. (2004). Balance as bias: global warming and the US prestige press. *Global Environmental Change–Human and Policy Dimensions, 14*(2), 125–136.

Boykoff, M. T., McNatt, M. M., & Goodman, M. K. (2015). Communicating in the anthropocene: the cultural politics of climate change news coverage around the world. In A. Hansen & R. Cox (Eds.), *The Routledge Handbook of Environment and Communication* (pp. 221–231). London and New York: Routledge.

Boykoff, M. T., & Yulsman, T. (2013). Political economy, media, and climate change: sinews of modern life. *Wiley Interdisciplinary Reviews–Climate Change, 4*(5), 359–371. doi: 10.1002/wcc.233

Boyle, D. (2017). Sierra Leone landslide: more than 300 killed as roads turn into 'churning rivers of mud'. *The Telegraph.* Retrieved 4 September, 2017, from http://www.telegraph.co.uk/news/2017/08/14/sierra-leone-landslide-hundreds-feared-buried-houses-submerged/

Brainard, C. (2015). The changing ecology of news and news organisations: implications for environmental news. In A. Hansen & R. Cox (Eds.), *The Routledge Handbook of Environment and Communication* (pp. 168–175). London and New York: Routledge.

Brosius, H.-B., & Kepplinger, H. M. (1990). The agenda-setting function of television news. *Communication Research, 17*(2), 183–211.

Brossard, D., Shanahan, J., & McComas, K. (2004). Are issue-cycles culturally constructed? A comparison of French and American coverage of global climate change. *Mass Communication & Society, 7*(3), 359–377. doi: DOI:10.1207/s15327825mcs0703_6

Bruggemann, M., & Engesser, S. (2014). Between consensus and denial: climate journalists as interpretive community. *Science Communication, 36*(4), 399–427. doi: 10.1177/1075547014533662

Brulle, R. J., Carmichael, J., & Jenkins, J. C. (2012). Shifting public opinion on climate change: an empirical assessment of factors influencing concern over climate change in the U.S., 2002–2010. *Climatic Change, 114*(2), 169–188. doi: 10.1007/s10584-012-0403-y

Buckley, R., & Vogt, S. (1996). Fact and emotion in environmental advertising by government, industry and community groups. *Ambio, 25*(3), 214–215.

Burgess, J., & Harrison, C. M. (1993). The circulation of claims in the cultural politics of environmental change. In A. Hansen (Ed.), *The Mass Media and Environmental Issues* (pp. 198–221). Leicester: Leicester University Press.

Calhoun, C. (Ed.). (2002). *Dictionary of the Social Sciences*. Oxford: Oxford University Press: Oxford Reference Online.

Campbell, F. (1999). *The Construction of Environmental News: A Study of Scottish Journalism*. Abingdon: Ashgate.

Cantrill, J. G., & Oravec, C. L. (Eds.). (1996). *The Symbolic Earth: Discourse and Our Creation of the Environment*. Lexington, KY: University of Kentucky Press.

Carmichael, J. T., & Brulle, R. J. (2017). Elite cues, media coverage, and public concern: an integrated path analysis of public opinion on climate change, 2001–2013. *Environmental Politics, 26*(2), 232–252. doi: 10.1080/09644016.2016.1263433

Carson, R. (1962). *Silent Spring*. Boston: Houghton Mifflin.

Chapman, G., Kumar, K., Fraser, C., & Gaber, I. (1997). *Environmentalism and the Mass Media: The North-South Divide*. London: Routledge.

Cho, B., Kwon, U., Gentry, J. W., Jun, S., & Kropp, F. (1999). Cultural values reflected in theme and execution: a comparative study of US and Korean television commercials. *Journal of Advertising, 28*(4), 59–73.

Clancy, M. (2011). Re-presenting Ireland: tourism, branding and national identity in Ireland. *Journal of International Relations and Development, 14*(3), 281–308. doi: 10.1057/jird.2010.4

Cobb, R. W., & Elder, C. D. (1971). The politics of agenda building: an alternative perspective for modern democratic theory. *Journal of Politics, 33*, 892–915.

Cohen, B. C. (1963). *The Press and Foreign Policy*. Princeton, NJ: Princeton University Press.

Coleman, R. R. M., & Yochim, E. C. (2008). Symbolic annihilation. In W. Donsbach (Ed.), *The Blackwell International Encyclopedia of Communication*. Oxford: Blackwell.
Collins, H. M. (1987). Certainty and the public understanding of science: science on television. *Social Studies of Science, 17*(4), 689–713.
Condit, C. M., Achter, P. J., Lauer, I., & Sefcovic, E. (2002). The changing meanings of "mutation": A contextualized study of public discourse. *Human Mutation, 19*(1), 69–75.
Conrad, P. (1999). Uses of expertise: sources, quotes, and voice in the reporting of genetics in the news. *Public Understanding of Science, 8*(4), 285–302.
Corbett, J. B. (1998). The environment as theme and package on a local television newscast. *Science Communication, 19*(3), 222–237.
Corbett, J. B. (2002). A faint green sell: advertising and the natural world. In M. Meister & P. M. Japp (Eds.), *Enviropop: Studies in Environmental Rhetoric and Popular Culture* (pp. 141–160). Westport, Conn.: Praeger/Greenwood Press.
Corbett, J. B. (2006). *Communicating Nature: How We Create and Understand Environmental Messages*. Washington, DC: Island Press.
Corbett, J. B., & Durfee, J. L. (2004). Testing public (un)certainty of science – Media representations of global warming. *Science Communication, 26*(2), 129–151. doi: 10.1177/1075547004270234
Corner, J., & Schlesinger, P. (1991). Editorial: covering the environment. *Media, Culture & Society, 13*(4), 435–441.
Cottle, S. (1993). Mediating the environment: modalities of TV news. In A. Hansen (Ed.), *The Mass Media and Environmental Issues* (pp. 107–133). Leicester: Leicester University Press.
Cottle, S. (1998). Ulrich Beck, 'risk society' and the media – a catastrophic view? *European Journal of Communication, 13*(1), 5–32.
Cottle, S. (2000). TV news, lay voices and the visualisation of environmental risks. In S. Allan, B. Adam, & C. Carter (Eds.), *Environmental Risks and the Media* (pp. 29–44). London: Routledge.
Cottle, S. (2004). Producing nature(s): on the changing production ecology of natural history TV. *Media Culture & Society, 26*(1), 81–101.
Cottle, S. (2006). *Mediatized Conflict: Developments in Media and Conflict Studies*. Maidenhead: Open University Press.
Cottle, S. (2009). *Global Crisis Reporting: Journalism in the Global Age*. Maidenhead: Open University Press.
Cox, R. (2006). *Environmental Communication and the Public Sphere*. London: Sage.
Cox, R., & Pezzullo, P. C. (2016). *Environmental Communication and the Public Sphere* (4th ed.). Thousand Oaks, California: SAGE Publications, Inc.
Cox, R., & Schwarze, S. (2015). The media/communication strategies of environmental pressure groups and NGOs. In A. Hansen & R. Cox (Eds.),

The Routledge Handbook of Environment and Communication (pp. 73–85). London and New York: Routledge.
Cracknell, J. (1993). Issue arenas, pressure groups and environmental agendas. In A. Hansen (Ed.), *The Mass Media and Environmental Issues* (pp. 3–21). Leicester: Leicester University Press.
Crawley, C. E. (2007). Localized debates of agricultural biotechnology in community newspapers – a quantitative content analysis of media frames and sources. *Science Communication, 28*(3), 314–346.
Creighton, M. (1997). Consuming rural Japan: the marketing of tradition and nostalgia in the Japanese travel industry. *Ethnology, 36*(3), 239–254.
Cronon, W. (Ed.). (1995). *Uncommon Ground: Toward Reinventing Nature.* New York: Norton.
Daley, P., & O'Neill, D. (1991). Sad is too mild a word – press coverage of the Exxon Valdez oil-spill. *Journal of Communication, 41*(4), 42–57.
Davies, G. (2000a). Narrating the Natural History Unit: institutional orderings and spatial strategies. *Geoforum, 31*(4), 539–551.
Davies, G. (2000b). Science, observation and entertainment: competing visions of postwar British natural history television, 1946–1967. *Ecumene, 7*(4), 432–460.
Davis, A. (2002). *Public Relations Democracy: Public Relations, Politics and the Mass Media in Britain.* Manchester: Manchester University Press.
Davis, A. (2003). Public relations and news sources. In S. Cottle (Ed.), *News, Public Relations and Power* (pp. 27–42). London: Sage.
Davis, A. (2013). *Promotional Cultures: The Rise and Spread of Advertising, Public Relations, Marketing and Branding.* Cambridge, UK: Polity.
Davis, F. (1979). *Yearning for Yesterday: A Sociology of Nostalgia.* New York: The Free Press.
de Jong, W. (2005). Limits and possibilities of media-based oppositional politics; Greenpeace versus Shell; the Brent Spar conflict. In W. de Jong, M. Shaw, & N. Stammers (Eds.), *Global Activism, Global Media* (pp. 110–124). London: Pluto.
Deacon, D., & Golding, P. (1993). Barriers to centralism – local government, local media and the charge on the community. *Local Government Studies, 19*(2), 176–189.
DEFRA. (2016). Guidance: make an environmental claim for your product, service or organisation. Retrieved 11 March, 2018, from https://www.gov.uk/government/publications/make-a-green-claim/make-an-environmental-claim-for-your-product-service-or-organisation#organisations-that-enforce-environmental-claims
DeLuca, K. M., & Peeples, J. (2002). From public sphere to public screen: democracy, activism, and the "violence" of Seattle. *Critical Studies in Media Communication, 19*(2), 125–151.
Deverson, T. (Ed.). (2004). *The New Zealand Oxford Dictionary.* Oxford: Oxford University Press: Oxford Reference Online.

Dingwall, R., & Aldridge, M. (2006). Television wildlife programming as a source of popular scientific information: a case study of evolution. *Public Understanding of Science, 15*(2), 131–152.

Djerf-Pierre, M. (2012a). When attention drives attention: issue dynamics in environmental news reporting over five decades. *European Journal of Communication, 27*(3), 291–304. doi: 10.1177/0267323112450820

Djerf-Pierre, M. (2012b). The crowding-out effect: issue dynamics and attention to environmental issues in television news reporting over 30 years. *Journalism Studies, 13*(4), 499–516. doi: 10.1080/1461670x.2011.650924

Djerf-Pierre, M. (2013). Green metacycles of attention: reassessing the attention cycles of environmental news reporting 1961–2010. *Public Understanding of Science, 22*(4), 495–512. doi: 10.1177/0963662511426819

Donohue, G. A., Olien, C. N., & Tichenor, P. J. (1989). Structure and constraints on community newspaper gatekeepers. *Journalism Quarterly, 66*, 807–812.

Donohue, G. A., Tichenor, P. J., & Olien, C. N. (1995). A guard dog perspective on the role of media. *Journal of Communication, 45*(2), 115–132.

Downs, A. (1972). Up and down with ecology – the issue attention cycle. *The Public Interest, 28*(3), 38–50.

Doyle, J. (2007). Picturing the clima(c)tic: Greenpeace and the representational politics of climate change communication. *Science as Culture, 16*(2), 129–150.

Doyle, J. (2011). *Mediating Climate Change*: Farnham, UK: Ashgate.

Dunlap, R. E. (1991). Trends in public opinion toward environmental-issues: 1965–1990. *Society & Natural Resources, 4*(3), 285–312.

Dunlap, R. E. (2006). Show us the data – the questionable empirical foundations of "The Death of Environmentalism" thesis. *Organization & Environment, 19*(1), 88–102.

Dunwoody, S. (1979). News-gathering behaviors of specialty reporters: a two-level comparison of mass media decision-making. *Newspaper Research Journal, 1*(1), 29–41.

Dunwoody, S. (1980). The science writing inner club: a communication link between science and the lay public. *Science, Technology, & Human Values, 5*, 14–22.

Dunwoody, S. (2014). Science journalism: prospects in the digital age. In M. Bucchi & B. Trench (Eds.), *Handbook of Public Communication of Science and Technology* (2nd ed., pp. 27–39). London: Routledge.

Dunwoody, S. (2015). Environmental scientists and public communication. In A. Hansen & R. Cox (Eds.), *The Routledge Handbook of Environment and Communication* (pp. 63–72). London and New York: Routledge.

Dunwoody, S., & Griffin, R. J. (1993). Journalistic strategies for reporting long-term environmental issues: a case study of three Superfund sites. In A. Hansen (Ed.), *The Mass Media and Environmental Issues* (pp. 22–50). Leicester: Leicester University Press.

Durant, J., Hansen, A., & Bauer, M. (1996). Public understanding of the new genetics. In M. Marteau & J. Richards (Eds.), *The Troubled Helix* (pp. 235–248). Cambridge: Cambridge University Press.

Edelman, M. (1988). *Constructing the Political Spectacle*. Chicago: University of Chicago Press.

Einsiedel, E., & Coughlan, E. (1993). The Canadian press and the environment: reconstructing a social reality. In A. Hansen (Ed.), *The Mass Media and Environmental Issues* (pp. 134–149). Leicester: Leicester University Press.

Elbro, C. (1983). *Det Overtalende Landskab: Ideer om Menneske og Samfund i Digternes og Annonceindustriens Naturskildringer i 1970'erne*. København: C.A. Reitzel.

Encyclopedia of Marxism. (2009). Commodification. Retrieved 18 March, 2018, from http://www.marxists.org/glossary/terms/c/o.htm#commodification

Entman, R. M. (1993). Framing: toward clarification of a fractured paradigm. *Journal of Communication, 43*(4), 51–58.

Entman, R. M. (2003). Cascading activation: contesting the White House's frame after 9/11. *Political Communication, 20*(4), 415–432.

Ericson, R. V., Baranek, P. M., & Chan, J. B. L. (1989). *Negotiating Control: A Study of News Sources*. Milton Keynes: Open University Press.

Eurobarometer. (2017). Attitudes of European citizens towards the environment: summary. European Commission.

Evernden, N. (1989). Nature in industrial society. In I. Angus & S. Jhally (Eds.), *Cultural Politics in Contemporary America* (pp. 151–164). New York: Routledge.

Eyerman, R., & Jamison, A. (1989). Environmental knowledge as an organizational weapon: the case of Greenpeace. *Social Science Information, 28*(1), 99–119.

Fairclough, N. (1989). *Language and Power*. London: Longman.

Fan, D. P., Brosius, H. B., & Kepplinger, H. M. (1994). Predictions of the public agenda from television coverage. *Journal of Broadcasting & Electronic Media, 38*(2), 163–177.

Farand, C. (2017). Floods in India, Bangladesh and Nepal kill 1,200 and leave millions homeless; authorities say monsoon flooding is one of the worst in region in years. *The Independent*. Retrieved 4 September, 2017, from http://www.independent.co.uk/news/world/asia/india-floods-bangladesh-nepal-deaths-millions-homeless-latest-news-updates-a7919006.html

Frandsen, F., & Johansen, W. (2017). Strategic communication. In C. R. Scott & L. K. Lewis (Eds.), *The International Encyclopedia of Organizational Communication* (pp. 2250–2258). Oxford, UK: Wiley-Blackwell.

Frewer, L. J. (2002). The media and genetically modified foods: evidence in support of social amplification of risk. *Risk Analysis, 22*(4), 701–711.

Friedman, S. (2004). And the beat goes on: the third decade of environmental journalism. In S. Senecah, S. Depoe, M. Neuzil, & G. Walker (Eds.),

The Environmental Communication Yearbook, Vol. 1 (Vol. 1, pp. 175–187). London: Lawrence Erlbaum Associates.

Friedman, S. (2015). The changing face of environmental journalism in the United States. In A. Hansen & R. Cox (Eds.), *The Routledge Handbook of Environment and Communication* (pp. 144–157). London and New York: Routledge.

Friedman, S. M. (1986). The journalist's world. In S. M. Friedman, S. Dunwoody, & C. L. Rogers (Eds.), *Scientists and Journalists: Reporting Science as News* (pp. 17–41). New York: The Free Press.

Friedman, S. M., Dunwoody, S., & Rogers, C. L. (Eds.). (1986). *Scientists and Journalists: Reporting Science as News*. New York: The Free Press.

Fuller, R., & Myers, R. (1941). The natural history of a social problem. *American Sociological Review, 6*(June), 320–328.

Funkhouser, G. R. (1973). The issues of the sixties: an exploratory study in the dynamics of public opinion. *Public Opinion Quarterly, 37*(1), 62–75.

Gaber, I. (2000). The greening of the public, politics and the press, 1985–1999. In J. Smith (Ed.), *The Daily Globe: Environmental Change, the Public and the Media* (pp. 115–126). London: Earthscan Publications.

Galtung, J., & Ruge, M. H. (1965). The structure of foreign news. *Journal of International Peace Research, 1*, 64–90.

Gamson, W. (1985). Goffman's legacy to political sociology. *Theory and Society, 14*(5), 605–622.

Gamson, W. A. (1988). A constructionist approach to mass media and public opinion. *Symbolic Interaction, 11*(2), 161–174.

Gamson, W. A., & Modigliani, A. (1989). Media discourse and public opinion on nuclear power: a constructionist approach. *American Journal of Sociology, 95*(1), 1–37.

Gandy, O. H. (1982). *Beyond Agenda Setting: Information Subsidies and Public Policy*. Norwood, NJ: Ablex Publishing.

Gans, H. J. (2004). *Deciding What's News*. New York: Northwestern University Press.

Gaskell, G., Bauer, M. W., Durant, J., & Allum, N. C. (1999). Worlds apart? The reception of genetically modified foods in Europe and the US. *Science, 285*(5426), 384–387.

Geller, G., Bernhardt, B. A., Gardner, M., Rodgers, J., & Holtzman, N. A. (2005). Scientists' and science writers' experiences reporting genetic discoveries: toward an ethic of trust in science journalism. *Genetics in Medicine, 7*(3), 198–205. doi: 10.1097/01.gim.0000156699.78856.23

Gerbner, G. (1972). Violence in television drama: trends and symbolic functions. In G. A. Comstock & E. A. Rubinstein (Eds.), *Media Content and Control: Television and Social Behavior* (Vol. 1, pp. 28–187). Washington, DC: U.S. Government Printing Office.

Gerbner, G., & Gross, L. (1976). Living with television: the violence profile. *Journal of Communication, 26*(2), 173–199.

Gerbner, G., Gross, L., Morgan, M., & Signorielli, N. (1994). Growing up with television: the cultivation perspective. In J. Bryant & D. Zimmerman (Eds.), *Media Effects: Advances in Theory and Research* (pp. 17–41). Hillsdale, NJ: Lawrence Erlbaum Associates.

Gitlin, T. (1980). *The Whole World is Watching: Mass Media in the Making and Unmaking of the New Left.* Berkeley: University of California Press.

Glasgow University Media Group. (1976). *Bad News.* London: Routledge & Kegan Paul.

Globescan. (2013). Environmental concerns "at record lows": global poll. Retrieved 5 February, 2018, from https://globescan.com/environmental-concerns-at-record-lows-global-poll/

Goldenberg, E. N. (1975). *Making the Papers: Access of Resource Poor Groups to the Metropolitan Press.* Lexington, MA: D.C. Heath.

Goldman, R., & Papson, S. (1996). *Sign Wars: The Cluttered Landscape of Advertising.* New York, NY: The Guilford Press.

Gooch, G. D. (1996). Environmental concern and the Swedish press – a case study of the effects of newspaper reporting, personal experience and social interaction on the public's perception of environmental risks. *European Journal of Communication, 11*(1), 107–127.

Good, J. E. (2009). The cultivation, mainstreaming, and cognitive processing of environmentalists watching television. *Environmental Communication – a Journal of Nature and Culture, 3*(3), 279–297. doi: 10.1080/17524030903229746

Good, J. E. (2013). *Television and the Earth: Not a Love Story.* Black Point, NS: Fernwood Publishing.

Goodell, R. (1987). The role of the mass media in scientific controversy. In H. T. Engelhardt & A. L. Caplan (Eds.), *Scientific Controversies* (pp. 585–597). Cambridge: Cambridge University Press.

Gore, A. (2006). *An Inconvenient Truth: The Planetary Emergency of Global Warming and What We Can Do About It.* London: Bloomsbury.

Grant, W. (2000). *Pressure Groups and British Politics* (2nd ed.). Basingstoke: Palgrave Macmillan.

Grunig, L. A., Grunig, J. E., & Dozier, D. M. (2002). *Excellent Public Relations and Effective Organizations: A Study of Communication Management in Three Countries.* Mahwah: Lawrence Erlbaum.

Guggenheim, D., & Gore, A. (2006). An Inconvenient Truth: A Global Warning [DVD]: Paramount.

Gutteling, J. M. (2005). Mazur's hypothesis on technology controversy and media. *International Journal of Public Opinion Research, 17*(1), 23–41.

Habermas, J. (1989). *The Structural Transformation of the Public Sphere: An Inquiry into a Category of Bourgeois Society.* Cambridge, MA: MIT Press.

Hackett, R. A. (2015). Climate crisis and communication: reflections on Naomi Klein's This Changes Everything. *Media and Communication, 3*(1), 1–4. doi: 10.17645/mac.v3i1.304

Hall, S. (1975). The 'structured communication' of events. In Unesco (Ed.), *Getting the Message Across* (pp. 115–145). Paris: The Unesco Press.

Hall, S. (1981). The determinations of news photographs. In S. Cohen & J. Young (Eds.), *The Manufacture of News* (Revised ed., pp. 226–243). London: Constable.

Hall, S. (1982). The rediscovery of 'ideology': return of the repressed in media studies. In M. Gurevitch, T. Bennett, J. Curran, & J. Woollacott (Eds.), *Culture, Society and the Media* (pp. 56–90). London: Methuen.

Hall, S., Critcher, C., Jefferson, T., Clarke, J., & Roberts, B. (1978). *Policing the Crisis*. London: Macmillan.

Halloran, J. D., Elliott, P., & Murdock, G. (1970). *Demonstrations and Communication*. Harmondsworth: Penguin.

Hannigan, J. A. (1995). *Environmental Sociology*. London: Routledge.

Hannigan, J. A. (2006). *Environmental Sociology* (2nd ed.). London: Routledge.

Hannigan, J. A. (2014). *Environmental Sociology* (3rd ed.). London: Routledge.

Hansen, A. (Ed.). (1993a). *The Mass Media and Environmental Issues*. Leicester: Leicester University Press.

Hansen, A. (1993b). Greenpeace and press coverage of environmental issues. In A. Hansen (Ed.), *The Mass Media and Environmental Issues* (pp. 150–178). Leicester: Leicester University Press.

Hansen, A. (1994). Journalistic practices and science reporting in the British press. *Public Understanding of Science, 3*(2), 111–134.

Hansen, A. (2000). Claimsmaking and framing in British newspaper coverage of the Brent Spar controversy. In S. Allan, B. Adam, & C. Carter (Eds.), *Environmental Risks and the Media* (pp. 55–72). London: Routledge.

Hansen, A. (2002). Discourses of nature in advertising. *Communications, 27*(4), 499–511. doi: DOI: 10.1515/comm.2002.005

Hansen, A. (2006). Tampering with nature: 'nature' and the 'natural' in media coverage of genetics and biotechnology. *Media, Culture & Society, 28*(6), 811–834. doi: 10.1177/0163443706067026

Hansen, A. (2010). *Environment, Media and Communication*. London: Routledge.

Hansen, A. (2011). Communication, media and environment: towards reconnecting research on the production, content and social implications of environmental communication. *International Communication Gazette, 73*(1–2), 7–25. doi: DOI: 10.1177/1748048510386739

Hansen, A. (Ed.). (2014). *Media and the Environment* (Vol. 1–4). London: Routledge.

Hansen, A. (2015a). News coverage of the environment: a longitudinal perspective. In A. Hansen & R. Cox (Eds.), *The Routledge Handbook of Environment and Communication* (pp. 209–220). London: Routledge.

Hansen, A. (2015b). Promising Directions for Environmental Communication Research. *Environmental Communication, 9*(3), 384–391. doi: 10.1080/17524032.2015.1044047

Hansen, A. (2016). The changing uses of accuracy in science communication. *Public Understanding of Science, 25*(7), 760–774. doi: 10.1177/0963662516636303

Hansen, A. (2017a). Media representation: environment. In P. Rössler (Ed.), *The International Encyclopedia of Media Effects*. Chichester, West Sussex; Malden, MA: John Wiley & Sons, Inc.

Hansen, A. (2017b). Methods for assessing visual images and depictions of climate change. *Oxford Research Encyclopedia of Climate Science*. Retrieved 27 Sep. 2017, from http://climatescience.oxfordre.com/view/10.1093/acrefore/9780190228620.001.0001/acrefore-9780190228620-e-491

Hansen, A. (2018). Using visual images for showing environmental problems. In A. Fill & H. Penz (Eds.), *The Routledge Handbook of Ecolinguistics* (pp. 179–195). London: Routledge.

Hansen, A., & Cox, R. (Eds.). (2015). *The Routledge Handbook of Environment and Communication*. London: Routledge.

Hansen, A., & Linné, O. (1994). Journalistic practices and television coverage of the environment: an international comparison. In C. Hamelink & O. Linné (Eds.), *Mass Communication Research: On Problems and Policies* (pp. 369–383). Norwood, New Jersey: Ablex.

Hansen, A., & Machin, D. (2008). Visually branding the environment: climate change as a marketing opportunity. *Discourse Studies, 10*(6), 777–794. doi: 10.1177/1461445608098200

Hansen, A., & Machin, D. (2013). Visually branding the environment: climate change as a marketing opportunity. In R. Wodak (Ed.), *Critical Discourse Analysis: SAGE Benchmarks in Language and Linguistics* (pp. 269–287). London: Sage.

Hargreaves, I., & Ferguson, G. (2000). *Who's Misunderstanding Whom? Bridging the Gulf of Understanding Between the Public, the Media and Science*. Swindon: ESRC.

Hargreaves, I., Lewis, J., & Speers, T. (2004). *Towards a Better Map: Science, the Public and the Media*. Swindon: ESRC.

Hart, P. S., Nisbet, E. C., & Myers, T. A. (2015). Public attention to science and political news and support for climate change mitigation. *Nature Climate Change, 5*(6), 541–545.

Hartmann, P. (1975). Industrial relations and the news media. *Industrial Relations Journal, 6*(4).

Healy, G., & Williams, P. (2017). Metaphor use in the political communication of major resource projects in Australia. *Pacific Journalism Review, 23*(1), 150–168. doi: 10.24135/pjr.v23i1.103

Heath, R. L., & Johansen, W. (Eds.). (2018). *The International Encyclopedia of Strategic Communication*. London: Wiley.

Hestres, L. E., & Hopke, J. E. (2017). Internet-enabled activism and climate change. *Oxford Research Encyclopedia of Climate Science*. Retrieved 2 Oct. 2017, from http://climatescience.oxfordre.com/view/10.1093/acrefore/9780190228620.001.0001/acrefore-9780190228620-e-404

Hilgartner, S. (1990). The dominant view of popularization – conceptual problems, political uses. *Social Studies of Science, 20*(3), 519–539.

Hilgartner, S., & Bosk, C. L. (1988). The rise and fall of social problems: a public arenas model. *American Journal of Sociology, 94*(1), 53–78.

Holbert, R. L., Kwak, N., & Shah, D. V. (2003). Environmental concern, patterns of television viewing, and pro-environmental behaviors: integrating models of media consumption and effects. *Journal of Broadcasting & Electronic Media, 47*(2), 177–196.

Hornig, S. (1990). Television's NOVA and the construction of scientific truth. *Critical Studies in Mass Communication, 7*(1), 11–23.

Howlett, M., & Raglon, R. (1992). Constructing the environmental spectacle: green advertisements and the greening of the corporate image. *Environmental History Review, 16*(4), 53–68.

Huxford, J. (2000). Framing the future: science fiction frames and the press coverage of cloning. *Continuum: Journal of Media and Cultural Studies, 14*(2), 187–199.

Ibarra, P. R., & Kitsuse, J. I. (1993). Vernacular constituents of moral discourse: an interactionist proposal for the study of social problems. In J. A. Holstein & G. Miller (Eds.), *Reconsidering Social Constructionism: Debates in Social Problems Theory* (pp. 25–58). Hawthorne, NY: Aldine de Gruyter.

IPCC. (2018). IPCC Factsheet: What is the IPCC? Retrieved 18 March, 2018, from https://www.ipcc.ch/news_and_events/docs/factsheets/FS_what_ipcc.pdf

Iyer, E., & Banerjee, B. (1993). Anatomy of green advertising. *Advances in Consumer Research, 20*, 494–501.

Jenner, E. (2012). News photographs and environmental agenda setting. *Policy Studies Journal, 40*, 274–301.

Kellert, S. R. (1995). Concepts of nature east and west. In M. E. Soule & G. Lease (Eds.), *Reinventing Nature: Responses to Postmodern Deconstruction* (pp. 103–122). Washington, DC: Island Press.

Kielbowicz, R. B., & Scherer, C. (1986). The role of the press in the dynamics of social movements. In G. Lang & K. Lang (Eds.), *Research in Social Movements, Conflicts and Change* (pp. 71–96). Greenwich, CT: JAI Press Inc.

Kilbourne, W. E. (1995). Green advertising – salvation or oxymoron. *Journal of Advertising, 24*(2), 7–19.

Kirby, D. A. (2014). Science and technology in film: themes and representations. In M. Bucchi & B. Trench (Eds.), *Handbook of Public Communication of Science and Technology* (2nd ed., pp. 97–112). London: Routledge.

Kitsuse, J. I., & Spector, M. (1973). Toward a sociology of social problems: social conditions, value judgments and social problems. *Social Problems, 20*(4), 407–419.

Klein, N. (2014). *This Changes Everything: Capitalism vs. the Climate*. London: Penguin.

Kloek, M. E., Elands, B. H. M., & Schouten, M. G. C. (2017). Race/ethnicity in visual imagery of Dutch nature conservation organizations. *Society & Natural Resources, 30*(9), 1033–1048. doi: 10.1080/08941920.2017.1295500

Knowles, E. (Ed.). (2006). *A Dictionary of Phrase and Fable*. Oxford: Oxford University Press.

Kress, G. (1997). Language in the media. In O. Boyd-Barrett (Ed.), *M.A. Mass Communications, Distance Learning* (Vol. Module 9, Unit 49, pp. 13–43). Leicester: University of Leicester.

Krieghbaum, H. (1967). *Science and the Mass Media*. New York: New York University Press.

Kristiansen, S. (2017). Characteristics of the mass media's coverage of nuclear energy and its risk: a literature review. *Sociology Compass, 11*(7), e12490-n/a. doi: 10.1111/soc4.12490

Kroma, M. M., & Flora, C. B. (2003). Greening pesticides: a historical analysis of the social construction of farm chemical advertisements. *Agriculture and Human Values, 20*(1), 21–35.

Lacey, C., & Longman, D. (1993). The press and public access to the environment and development debate. *Sociological Review, 41*(2), 207–243.

Lacey, C., & Longman, D. (1997). *The Press as Public Educator: Cultures of Understanding, Cultures of Ignorance*. Luton: University of Luton Press.

LaFollette, M. C. (1990). *Making Science Our Own: Public Images of Science 1910–1955*. Chicago: The University of Chicago Press.

Lahtinen, R., & Vuorisalo, T. (2005). In search for the roots of environmental concern – water management and animal welfare issues in the Finnish local press in 1890–1950. *Scandinavian Journal of History, 30*(2), 177–197.

Lester, L. (2010). *Media and Environment*. Cambridge: Polity.

Lester, L., & Cottle, S. (2009). Visualizing climate change: television news and ecological citizenship. *International Journal of Communication, 3*, 920–936.

Lester, L., & Cottle, S. (2015). Transnational protests, publics and media participation (in an environmental age). In A. Hansen & R. Cox (Eds.), *The Routledge Handbook of Environment and Communication* (pp. 100–110). London and New York: Routledge.

Lewandowsky, S., Ecker, U. K. H., & Cook, J. (2017). Beyond misinformation: understanding and coping with the "post-truth" era. *Journal of Applied Research in Memory and Cognition, 6*(4), 353–369.

Lewis, J., Williams, A., & Franklin, B. (2008). A compromised fourth estate? UK news journalism, public relations and news sources. *Journalism Studies, 9*(1), 1–20.

Limoges, C. (1993). Expert knowledge and decision-making in controversy contexts. *Public Understanding of Science, 2*, 417–426.

Lin, C. A. (2001). Cultural values reflected in Chinese and American television advertising. *Journal of Advertising, 30*(4), 83–94.

Linder, S. H. (2006). Cashing-in on risk claims: on the for-profit inversion of signifiers for "global warming". *Social Semiotics, 16*(1), 103–132.

Linné, O., & Hansen, A. (1990). *News Coverage of the Environment: A Comparative Study of Journalistic Practices and Television Presentation in Denmark's Radio and the BBC*. Copenhagen: Danmarks Radio Forlaget.

Listerman, T. (2010). Framing of science issues in opinion-leading news: international comparison of biotechnology issue coverage. *Public Understanding of Science, 19*(1), 5–15. doi: 10.1177/0963662505089539

Liu, X., Lindquist, E., & Vedlitz, A. (2011). Explaining media and congressional attention to global climate change, 1969–2005: An empirical test of agenda-setting theory. *Political Research Quarterly, 64*(2), 405–419. doi: 10.1177/1065912909346744

Lloyd, J., & Toogood, L. (2015). *Journalism and PR: News Media and Public Relations in the Digital Age.* London: I. B. Tauris.

Lomborg, B. (2007). *Cool It: The Skeptical Environmentalist's Guide to Global Warming.* London: Marshall Cavendish: Cyan Communications Ltd.

Lowe, P., & Morrison, D. (1984). Bad news or good news: environmental politics and the mass media. *The Sociological Review, 32*(1), 75–90.

Lowe, P. D., & Rüdig, W. (1986). Review article: political ecology and the social sciences – the state of the art. *British Journal of Political Science, 16*, 513–550.

Lukes, S. (1974). *Power: A Radical View.* London and Basingstoke: Macmillan.

Lukes, S. (2005). *Power: A Radical View* (2nd edition). London and Basingstoke: Palgrave Macmillan.

Macnaghten, P., & Urry, J. (1998). *Contested Natures.* London: Sage.

Marchand, R. (1985). *Advertising the American Dream: Making Way for Modernity, 1920–1940.* Berkeley: University of California Press.

Martin, D. C. (2004). Apartheid in the great outdoors: American advertising and the reproduction of a racialized outdoor leisure identity. *Journal of Leisure Research, 36*(4), 513–535.

Mazur, A. (1981). Media coverage and public opinion on scientific controversies. *Journal of Communication, 31*(2), 106–115.

Mazur, A. (1984). The journalists and technology: reporting about Love Canal and Three Mile Island. *Minerva, 22*(Spring), 45–66.

Mazur, A. (1990). Nuclear power, chemical hazards, and the quantity of reporting. *Minerva, 28*(3), 294–323.

Mazur, A. (2016). How did the fracking controversy emerge in the period 2010–2012? *Public Understanding of Science, 25*(2), 207–222. doi: 10.1177/0963662514545311

Mazur, A., & Lee, J. (1993). Sounding the global alarm: environmental issues in the US national news. *Social Studies of Science, 23*(4), 681–720.

McComas, K., & Shanahan, J. (1999). Telling stories about global climate change – measuring the impact of narratives on issue cycles. *Communication Research, 26*(1), 30–57.

McComas, K. A., Shanahan, J., & Butler, J. S. (2001). Environmental content in prime-time network TV's non-news entertainment and fictional programs. *Society & Natural Resources, 14*(6), 533–542.

McCombs, M. (2014). *Setting the Agenda: The Mass Media and Public Opinion* (2nd ed.). Cambridge, UK: Polity Press.

McCombs, M., Holbert, R. L., Kiousis, S., & Wanta, W. (2011). *The News and Public Opinion: Media Effects on Civic Life*. Wiley.

McCombs, M. E., & Shaw, D. L. (1972). The agenda-setting function of mass media. *Public Opinion Quarterly, 36*, 176–187.

McQuail, D. (2010). *McQuail's Mass Communication Theory* (6th ed.). London: Sage.

Mikami, S., Takeshita, T., Kawabata, M., Sekiya, N., Nakada, M., Otani, N., & Takahashi, N. (2002). *Unsolved Conflict among Europe, Japan and USA on the Global Warming Issue: Analysis of the Longitudinal Trends in News Frame.* Paper presented at the IAMCR, Barcelona, Spain.

Mikami, S., Takeshita, T., Nakada, M., & Kawabata, M. (1995). The media coverage and public awareness of environmental issues in Japan. *International Communication Gazette, 54*(3), 209–226. doi: 10.1177/001654929505400302

Miller, A., & LaPoe, V. (2016). Visual agenda-setting, emotion, and the BP oil disaster. *Visual Communication Quarterly, 23*(1), 53–63. doi: 10.1080/15551393.2015.1128335

Miller, D. (1999). Risk, science and policy: definitional struggles, information management, the media and BSE. *Social Science and Medicine, 49*(9), 1239–1255.

Miller, D., & Dinan, W. (2015). Resisting meaningful action on climate change: think tanks, 'merchants of doubt' and the 'corporate capture' of sustainable development. In A. Hansen & R. Cox (Eds.), *The Routledge Handbook of Environment and Communication* (pp. 86–99). London and New York: Routledge.

Miller, G., & Holstein, J. A. (1993). Reconsidering social constructionism. In J. A. Holstein & G. Miller (Eds.), *Reconsidering Social Constructionism: Debates in Social Problems Theory* (pp. 5–23). Hawthorne, NY: Aldine deGruyter.

Miller, M. M., & Riechert, B. P. (2000). Interest group strategies and journalistic norms: news media framing of environmental issues. In S. Allan, B. Adam, & C. Carter (Eds.), *Environmental Risks and the Media* (pp. 45–54). London: Routledge.

Mills, C. W. (1940). Situated actions and vocabularies of motives. *American Sociological Review, 6*, 904–913.

Mitman, G. (1999). *Reel Nature: America's Romance with Wildlife on Film.* Cambridge, MA: Harvard University Press.

Molotch, H., & Lester, M. (1974). News as purposive behaviour. *American Sociological Review, 39*, 101–112.

Moon, Y. S., & Chan, K. (2005). Advertising appeals and cultural values in television commercials – a comparison of Hong Kong and Korea. *International Marketing Review, 22*(1), 48–66.

Moore, E. E. (2015). Green screen or smokescreen? Hollywood's messages about nature and the environment. *Environmental Communication*, 1–17. doi: 10.1080/17524032.2015.1014391

Moore, E. E. (2017). *Landscape and the Environment in Hollywood Film: The Green Machine*. Basingstoke and New York: Palgrave Macmillan.
Mueller, B. (1987). Reflections of culture – an analysis of Japanese and American advertising appeals. *Journal of Advertising Research, 27*(3), 51–59.
Murrell, R. K. (1987). Telling it like it isn't: representations of science in Tomorrow's World. *Theory, Culture & Society, 4*, 89–106.
Negra, D. (2001). Consuming Ireland: Lucky Charms cereal, Irish Spring soap and 1–800-SHAMROCK. *Cultural Studies, 15*(1), 76–97.
Nelkin, D. (1987). *Selling Science: How the Press Covers Science and Technology*. New York: W H Freeman & Company.
Nelkin, D. (1995). *Selling Science: How the Press Covers Science and Technology* (2nd Revised ed.). New York: W.H. Freeman.
Nelson, V. (2005). Representation and images of people, place and nature in Grenada's tourism. *Geografiska Annaler Series B-Human Geography, 87B*(2), 131–143.
Nerlich, B. (2003). Tracking the fate of the metaphor silent spring in British environmental discourse: towards an evolutionary ecology of metaphor. *Metaphorik.de, 4*, 115–140.
Nerlich, B. (2010). 'Climategate': paradoxical metaphors and political paralysis. *Environmental Values, 19*(4), 419–442. doi: 10.3197/096327110x531543
Nimmo, D., & Combs, J. E. (1981). 'The horror tonight': network television news and Three Mile Island. *Journal of Broadcasting, 25*(3), 289–293.
Nisbet, E. C., Cooper, K. E., & Ellithorpe, M. (2015). Ignorance or bias? Evaluating the ideological and informational drivers of communication gaps about climate change. *Public Understanding of Science, 24*(3), 285–301. doi: 10.1177/0963662514545909
Nisbet, M. (2009). Communicating climate change: why frames matter for public engagement. *Environment: Science and Policy for Sustainable Development*. Retrieved 3 February, March–April, from http://www.environmentmagazine.org/March-April%202009/Nisbet-full.html
Nisbet, M. (Ed.). (2018). *The Oxford Encyclopedia of Climate Change Communication*. Oxford, UK: Oxford University Press.
Nisbet, M. C., & Huge, M. (2006). Attention cycles and frames in the plant biotechnology debate – managing power and participation through the press/policy connection. *Harvard International Journal of Press/Politics, 11*(2), 3–40.
Nisbet, M. C., & Lewenstein, B. V. (2002). Biotechnology and the American media – The policy process and the elite press, 1970 to 1999. *Science Communication, 23*(4), 359–391.
Nisbet, M. C., & Newman, T. P. (2015). Framing, the media, and environmental communication. In A. Hansen & R. Cox (Eds.), *The Routledge Handbook of Environment and Communication* (pp. 325–338). London and New York: Routledge.
Norton, T., & Grecu, N. (2015). Publics, communication campaigns and persuasive communication. In A. Hansen & R. Cox (Eds.), *The Routledge*

Handbook of Environment and Communication (pp. 354–367). London and New York: Routledge.
Okazaki, S., & Mueller, B. (2008). Evolution in the usage of localised appeals in Japanese and American print advertising. *International Journal of Advertising, 27*(5), 771–798. doi: 10.2501/s0265048708080323
Oreskes, N. (2004). The scientific consensus on climate change. *Science, 306*(5720), 1686.
Oreskes, N., & Conway, E. M. (2010). *Merchants of Doubt: How a Handful of Scientists Obscured the Truth on Issues from Tobacco Smoke to Global Warming.* London: Bloomsbury.
Ott, B. L. (2017). The age of Twitter: Donald J. Trump and the politics of debasement. *Critical Studies in Media Communication, 34*(1), 59–68. doi: 10.1080/15295036.2016.1266686
Patterson, P. (1989). Reporting Chernobyl: cutting the government fog to cover the nuclear cloud. In L. M. Walters, L. Wilkins, & T. Walters (Eds.), *Bad Tidings: Communication and Catastrophe* (pp. 131–147). Hillsdale, NJ: Lawrence Erlbaum Associates.
Peeples, J. (2013). Imaging toxins. *Environmental Communication – a Journal of Nature and Culture, 7*(2), 191–210. doi: 10.1080/17524032.2013.775172
Petersen, A. (2001). Biofantasies: genetics and medicine in the print news media. *Social Science & Medicine, 52*(8), 1255–1268.
Peterson, R. T. (1991). Physical environment television advertisement themes: 1979 and 1989. *Journal of Business Ethics, 10*(3), 221–228.
Pew Research Center. (2017). Globally, people point to ISIS and climate change as leading security threats. *Pew Global Atitudes Survey.* from http://www.pewglobal.org/2017/08/01/globally-people-point-to-isis-and-climate-change-as-leading-security-threats/
Pezzullo, P. C., & Cox, R. (2018). *Environmental Communication and the Public Sphere* (5th ed.). London: Sage.
Phillips, M., Fish, R., & Agg, J. (2001). Putting together ruralities: towards a symbolic analysis of rurality in the British mass media. *Journal of Rural Studies, 17*(1), 1–27.
Philo, G., & Happer, C. (2013). *Communicating Climate Change and Energy Security: New Methods in Understanding Audiences.* London: Routledge.
Pinto, J., Prada, P., & Tirado-Alcaraz, J. A. (2017). *Environmental News in Latin America: Conflict, Crisis and Contestation.* Basingstoke and New York: Palgrave Macmillan.
Plec, E., & Pettenger, M. (2012). Greenwashing consumption: the didactic framing of ExxonMobil's energy solutions. *Environmental Communication – a Journal of Nature and Culture, 6*(4), 459–476. doi: 10.1080/17524032.2012.720270
Poberezhskaya, M. (2014). Media coverage of climate change in Russia: governmental bias and climate silence. *Public Understanding of Science, 24*(1), 96–111. doi: 10.1177/0963662513517848
Podeschi, C. (2002). The nature of future myths: environmental discourse in science fiction film, 1950–1999. *Sociological Spectrum, 22*(3), 251–297.

Pollach, I. (2018). Issue cycles in corporate sustainability reporting: a longitudinal study. *Environmental Communication – a Journal of Nature and Culture, 12*(2), 247–260. doi: 10.1080/17524032.2016.1205645

Porritt, J., & Winner, D. (1988). *The Coming of the Greens.* London: Fontana.

Prevot-Julliard, A. C., Julliard, R., & Clayton, S. (2015). Historical evidence for nature disconnection in a 70-year time series of Disney animated films. *Public Understanding of Science, 24*(6), 672–680. doi: 10.1177/0963662513519042

Priest, S. (2015). Mapping media's role in environmental thought and action. In A. Hansen & R. Cox (Eds.), *The Routledge Handbook of Environment and Communication* (pp. 301–311). London and New York: Routledge.

Priest, S. (2016). *Communicating Climate Change: The Path Forward.* Basingstoke and New York: Palgrave Macmillan.

Propp, V. (1968). *Morphology of the Folktale.* Austin, TX: University of Texas Press.

Rebich-Hespanha, S., Rice, R. E., Montello, D. R., Retzloff, S., Tien, S., & Hespanha, J. P. (2015). Image themes and frames in US print news stories about climate change. *Environmental Communication, 9*(4), 491–519. doi: 10.1080/17524032.2014.983534

Reese, S. D. (2001). Prologue – framing public life: a bridging model for media research. In S. D. Reese, O. H. Gandy, & A. E. Grant (Eds.), *Framing Public Life: Perspectives on Media and Our Understanding of the Social World* (pp. 7–31). Mahwah, NJ: Lawrence Erlbaum Associates.

Renzi, B. G., Cotton, M., Napolitano, G., & Barkemeyer, R. (2016). Rebirth, devastation and sickness: analyzing the role of metaphor in media discourses of nuclear power. *Environmental Communication*, 1–17. doi: 10.1080/17524032.2016.1157506

Rose, C. (2011). *How to Win Campaigns: Communications for Change* (2nd ed.). London: Routledge.

Roser-Renouf, C., Stenhouse, N., Rolfe-Redding, J., Maibach, E., & Leiserowitz, A. (2015). Engaging diverse audiences with climate change: message strategies for global warming's six Americas. In A. Hansen & R. Cox (Eds.), *The Routledge Handbook of Environment and Communication* (pp. 368–386). London and New York: Routledge.

Rothman, S., & Lichter, S. R. (1987). Elite ideology and risk perception in nuclear energy policy. *American Political Science Review, 81*(2), 383–404.

Rutherford, P. (2000). *Endless Propaganda: The Advertising of Public Goods.* Toronto: University of Toronto Press.

Ryan, C. (1991). *Prime Time Activism: Media Strategies for Grassroots Organizing.* Boston, MA: South End Press.

Sachsman, D. B. (1973). *Public relations influence on environmental coverage (in the San Francisco Bay area).* (Doctoral dissertation), Stanford University.

Sachsman, D. B. (1976). Public relations influence on coverage of environment in San Francisco Area. *Journalism Quarterly, 53*(1), 54–60.

Sachsman, D. B., Simon, J., & Valenti, J. M. (2006). Regional issues, national norms: a four-region analysis of U.S. environment reporters. *Science Communication, 28*(1), 93–121.

Sachsman, D. B., Simon, J., & Valenti, J. M. (2010). *Environment Reporters in the 21st Century*. New Brunswick, NJ: Transaction Publishers.

Sachsman, D. B., & Valenti, J. M. (2015). Environmental reporters. In A. Hansen & R. Cox (Eds.), *The Routledge Handbook of Environment and Communication*. London and New York: Routledge.

Sachsman, D. B., & Valenti, J. M. (Eds.). (2019). *The Routledge Handbook of Environmental Journalism*. London: Routledge.

Sahlins, M. (1977). *The Use and Abuse of Biology*. London: Tavistock Publishers.

Sampei, Y., & Aoyagi-Usui, M. (2009). Mass-media coverage, its influence on public awareness of climate-change issues, and implications for Japan's national campaign to reduce greenhouse gas emissions. *Global Environmental Change – Human and Policy Dimensions, 19*(2), 203–212. doi: 10.1016/j.gloenvcha.2008.10.005

San Deogracias, J. C., & Mateos-Pérez, J. (2013). Thinking about television audiences: entertainment and reconstruction in nature documentaries. *European Journal of Communication, 28*(5), 570–583. doi: 10.1177/0267323113494075

Schäfer, M. S., Ivanova, A., & Schmidt, A. (2014). What drives media attention for climate change? Explaining issue attention in Australian, German and Indian print media from 1996 to 2010. *International Communication Gazette, 76*(2), 152–176. doi: 10.1177/1748048513504169

Schama, S. (1995). *Landscape and Memory*. London: HarperCollins.

Schlesinger, P. (1990). Rethinking the sociology of journalism: source strategies and the limits of media centrism. In M. Ferguson (Ed.), *Public Communication: The New Imperatives* (pp. 61–83). London: Sage.

Schlesinger, P., & Tumber, H. (1994). *Reporting Crime: The Media Politics of Criminal Justice*. Oxford: Clarendon Press.

Schlichting, I. (2013). Strategic framing of climate change by industry actors: a meta-analysis. *Environmental Communication, 7*(4), 493–511. doi: 10.1080/17524032.2013.812974

Schmuck, D., Matthes, J., Naderer, B., & Beaufort, M. (2018). The effects of environmental brand attributes and nature imagery in green advertising. *Environmental Communication, 12*(3), 414–429. doi: 10.1080/17524032.2017.1308401

Schneider, J., Schwarze, S., Bsumek, P. K., & Peeples, J. (2016). *Under Pressure: Coal Industry Rhetoric and Neoliberalism*. Basingstoke and New York: Palgrave Macmillan.

Schneider, J. W. (1985). Social problems theory: the constructionist view. *Annual Review of Sociology, 11*, 209–229.

Schoenfeld, A. C. (1980). Newspersons and the environment today. *Journalism Quarterly, 57*, 456–462.

Schoenfeld, A. C., Meier, R. F., & Griffin, R. J. (1979). Constructing a social problem – the press and the environment. *Social Problems, 27*(1), 38–61. doi: 10.1525/sp.1979.27.1.03a00040

Schudson, M. (1989). The sociology of news production. *Media, Culture & Society, 11*(3), 263–282.

Schultz, F., Kleinnijenhuis, J., Oegema, D., Utz, S., & van Atteveldt, W. (2012). Strategic framing in the BP crisis: a semantic network analysis of associative frames. *Public Relations Review, 38*(1), 97–107.

Scott, J., & Marshall, G. (2009). *A Dictionary of Sociology*. Oxford: Oxford University Press.

Scott, K. D. (2003). Popularizing science and nature programming – the role of "Spectacle" in contemporary wildlife documentary. *Journal of Popular Film and Television, 31*(1), 29–35.

Scutt, R., & Bonnet, A. (1996). *In Search of England: Popular Representations of Englishness and the English Countryside* (pp. 33). Newcastle upon Tyne: University of Newcastle upon Tyne, Department of Agricultural Economics and Food Marketing; Centre for Rural Economy.

Segerberg, A. (2017). Online and social media campaigns for climate change engagement. *Oxford Research Encyclopedia of Climate Science*. Retrieved 23 Aug. 2017, from http://climatescience.oxfordre.com/view/10.1093/acrefore/9780190228620.001.0001/acrefore-9780190228620-e-398

Segev, S., Fernandes, J., & Hong, C. (2016). Is your product really green? a content analysis to reassess green advertising. *Journal of Advertising, 45*(1), 85–93. doi: 10.1080/00913367.2015.1083918

Shanahan, J. (1993). Television and the cultivation of environmental concern: 1988–92. In A. Hansen (Ed.), *The Mass Media and Environmental Issues* (pp. 181–197). Leicester: Leicester University Press.

Shanahan, J. (1996). Green but unseen: marginalizing the environment on television. In M. Morgan & S. Leggett (Eds.), *Mainstream(s) and Margins: Cultural Politics in the 90s*. Westport CT: Greenwood.

Shanahan, J., & McComas, K. (1997). Television's portrayal of the environment: 1991–1995. *Journalism & Mass Communication Quarterly, 74*(1), 147–159.

Shanahan, J., & McComas, K. (1999). *Nature Stories: Depictions of the Environment and their Effects*. Cresskill, N.J.: Hampton Press.

Shanahan, J., McComas, K., & Deline, M. B. (2015). Representations of the environment on television, and their effects. In A. Hansen & R. Cox (Eds.), *The Routledge Handbook of Environment and Communication*. London and New York: Routledge.

Shanahan, J., Morgan, M., & Stenbjerre, M. (1997). Green or brown? Television and the cultivation of environmental concern. *Journal of Broadcasting & Electronic Media, 41*(3), 305–323.

Shehata, A., & Hopmann, D. N. (2012). Framing climate change: a study of US and Swedish press coverage of global warming. *Journalism Studies, 13*(2), 175–192. doi: 10.1080/1461670x.2011.646396

Signitzer, B., & Prexl, A. (2007, 23–25 July). *Communication strategies of 'greenwash trackers' – how activist groups attempt to hold companies accountable and to promote sustainable development*. Paper presented at the IAMCR Annual Conference, Paris.

Smith, C. (1992). *Media and Apocalypse: News Coverage of the Yellowstone Forest Fires, Exxon Valdez Oil Spill, and Loma Prieta Earthquake.* Westport, CT: Greenwood Press.

Smith, C. (1996). Reporters, news sources, and scientific intervention: the new Madrid earthquake prediction. *Public Understanding of Science, 5*(3), 205–216.

Soanes, C., & Stevenson, A. (Eds.). (2005). *The Oxford Dictionary of English (revised edition).* Oxford: Oxford University Press: Oxford Reference Online.

Soanes, C., & Stevenson, A. (Eds.). (2008). *The Concise Oxford English Dictionary* (Twelfth edition ed.). Oxford: Oxford University Press: Oxford Reference Online.

Sobieraj, S., & Berry, J. M. (2011). From incivility to outrage: political discourse in blogs, talk radio, and cable news. *Political Communication, 28*(1), 19–41. doi: 10.1080/10584609.2010.542360

Solesbury, W. (1976). The environmental agenda: an illustration of how situations may become political issues and issues may demand responses from government; or how they may not. *Public Administration, 54*(4), 379–397.

Soper, K. (1995). *What is Nature?* Oxford: Blackwell.

Soroka, S. N. (2002). Issue attributes and agenda-setting by media, the public, and policymakers in Canada. *International Journal of Public Opinion Research, 14*(3), 264–285.

Sourcewatch. (2017). Greenwashing. Retrieved 18 March, 2018, from https://www.sourcewatch.org/index.php/Greenwashing

Spector, M., & Kitsuse, J. I. (1973). Social problems: a reformulation. *Social Problems, 21*(2), 145–159.

Spector, M., & Kitsuse, J. I. (1977). *Constructing Social Problems.* Menlo Park, CA: Cummings.

Spector, M., & Kitsuse, J. I. (1987). *Constructing Social Problems.* New York: Aldine de Gruyter.

Spector, M., & Kitsuse, J. I. (2000). *Constructing Social Problems* (New edition ed.). New Brunswick, NJ: Transaction Publishers.

Stallings, R. A. (1990). Media discourse and the social construction of risk. *Social Problems, 37*(1), 80–95.

Stallings, R. A. (1995). *Promoting Risk: Constructing the Earthquake Threat.* New York: Aldine De Gruyter.

Stocking, S. H. (1999). How journalists deal with scientific uncertainty. In S. M. Friedman, S. Dunwoody, & C. L. Rogers (Eds.), *Communicating Uncertainty: Media Coverage of New and Controversial Science* (pp. 23–42). Mahwah, NJ: Lawrence Erlbaum Associates.

Stöckl, H., & Molnar, S. (2018). Eco-advertising: the linguistics and semiotics of green(-washed) persuasion. In A. F. Fill & H. Penz (Eds.), *The Routledge Handbook of Ecolinguistics* (pp. 261–276). London: Routledge.

Svoboda, M. (2016). Cli-fi on the screen(s): patterns in the representations of climate change in fictional films. *Wiley Interdisciplinary Reviews – Climate Change, 7*(1), 43–64. doi: 10.1002/wcc.381

Swidler, A. (1986). Culture in action: symbols and strategies. *American Sociological Review, 51*(2), 273–286.

Taylor, C. E., Lee, J. S., & Davie, W. R. (2000). Local press coverage of environmental conflict. *Journalism & Mass Communication Quarterly, 77*(1), 175–192.

Ten Eyck, T. A., & Williment, M. (2003). The national media and things genetic – coverage in the New York Times (1971–2001) and the Washington Post (1977–2001). *Science Communication, 25*(2), 129–152.

Thomas, L. (1995). In love with Inspector Morse – feminist subculture and quality television. *Feminist Review* (51), 1–25.

Thomas, L. (2002). *Fans, Feminisms and 'Quality' Media*. London: Routledge.

Thompson, J. B. (1990). *Ideology and Modern Culture: Critical Social Theory in the Era of Mass Communication*. Oxford: Polity Press.

Tichenor, P. J., Donohue, G. A., & Olien, C. N. (1980). *Community Conflict and the Press*. Beverly Hills, CA: Sage.

Tong, J. (2014). Environmental risks in newspaper coverage: a framing analysis of investigative reports on environmental problems in 10 Chinese newspapers. *Environmental Communication*, 1–23. doi: 10.1080/17524032.2014.898675

Tong, J. (2015). *Investigative Journalism, Environmental Problems and Modernisation in China*. Basingstoke and New York: Palgrave Macmillan.

Trench, B. (2009). Science reporting in the electronic embrace of the internet. In R. Holliman, E. Whitelegg, E. Scanlon, S. Smidt, & J. Thomas (Eds.), *Investigating Science Communication in the Information Age: Implications for Public Engagement and Popular Media* (pp. 166–180). Milton Keynes: Oxford University Press and The Open University.

Trumbo, C. (1995). *Longitudinal Modelling of Public Issues: An Application of the Agenda-Setting Process to the Issue of Global Warming* (Vol. 152). Columbia, S.C.: Association for Education in Journalism and Mass Communication.

Trumbo, C. (1996). Constructing climate change: claims and frames in US news coverage of an environmental issue. *Public Understanding of Science, 5*(3), 269–283.

Trumbo, C., & Kim, S.-J. S. (2015). Agenda-setting with environmental issues. In A. Hansen & R. Cox (Eds.), *The Routledge Handbook of Environment and Communication* (pp. 312–324). London and New York: Routledge.

Tuchman, G. (1978). The symbolic annihilation of women by the mass media. In G. Tuchman, A. Kaplan Daniels, & J. Benet (Eds.), *Hearth and Home: Images of Women in the Mass Media* (pp. 3–17). New York: Oxford University Press.

Turney, J. (1998). *Frankenstein's Footsteps: Science, Genetics and Popular Culture*. London: Yale University Press.

Uggla, Y., & Olausson, U. (2013). The enrollment of nature in tourist information: framing urban nature as "the other". *Environmental Communication – a Journal of Nature and Culture, 7*(1), 97–112. doi: 10.1080/17524032.2012.745009

Ungar, S. (1992). The rise and (relative) decline of global warming as a social problem. *Sociological Quarterly, 33*(4), 483–501.

Ungar, S. (2003). Global warming versus ozone depletion: failure and success in North America. *Climate Research, 23*(3), 263–274.

Urry, J. (1995). *Consuming Places*. London: Sage.

Urry, J. (2001). *The Tourist Gaze: Leisure and Travel in Contemporary Societies* (2nd ed.). London: Sage.

Uscinski, J. E. (2009). When does the public's issue agenda affect the media's issue agenda (and vice-versa)? Developing a framework for media-public influence. *Social Science Quarterly, 90*(4), 796–815.

Väliverronen, E., & Hellsten, I. (2002). From "burning library" to "green medicine" – The role of metaphors in communicating biodiversity. *Science Communication, 24*(2), 229–245.

Wakefield, S. E. L., & Elliott, S. J. (2003). Constructing the news: the role of local newspapers in environmental risk communication. *Professional Geographer, 55*(2), 216–226.

Walgrave, S., Soroka, S., & Nuytemans, M. (2008). The mass media's political agenda-setting power – a longitudinal analysis of media, parliament, and government in Belgium (1993 to 2000). *Comparative Political Studies, 41*(6), 814–836. doi: 10.1177/0010414006299098

Wall, G. (1999). Science, nature, and The Nature of Things: an instance of Canadian environmental discourse, 1960–1994. *Canadian Journal of Sociology – Cahiers Canadiens De Sociologie, 24*(1), 53–85.

Wall, M. (2002). The battle in Seattle – how nongovernmental organizations used websites in their challenge to the WTO. In E. Gilboa (Ed.), *Media and Conflict: Framing Issues, Making Policy, Shaping Opinions* (pp. 25–43). Ardsley, NY: Transnational Publishers.

Wallack, L., Woodruff, K., Dorfman, L., & Diaz, I. (1999). *News for a Change: An Advocate's Guide to Working with the Media*. London: Sage.

Walters, L. M., Wilkins, L., & Walters, T. (Eds.). (1989). *Bad Tidings: Communication and Catastrophe*. Hillsdale, New Jersey: Lawrence Erlbaum Associates.

Watson, J., & Hill, A. (2015). *Dictionary of Media and Communication Studies* (9th ed.). London: Bloomsbury.

Weart, S. R. (1988). *Nuclear Fear: A History of Images*. Cambridge, MA: Harvard University Press.

Weart, S. R. (2012). *The Rise of Nuclear Fear*. Cambridge, MA: Harvard University Press.

Weber, M. (1930). *The Protestant Ethic and the Spirit of Capitalism*. London: Unwin University Books.

Wernick, A. (1997). Resort to nostalgia: mountains, memories and myths of time. In M. Nava (Ed.), *Buy this Book: Studies in Advertising and Consumption* (pp. 207–223). London: Routledge.

Whitmarsh, L. (2015). Analysing public perceptions, understanding and images of environmental change. In A. Hansen & R. Cox (Eds.), *The Routledge Handbook of Environment and Communication* (pp. 339–353). London and New York: Routledge.

Widener, P., & Gunter, V. J. (2007). Oil spill recovery in the media: missing an Alaska native perspective. *Society & Natural Resources, 20*, 767–783.

Wiegman, O., Gutteling, J. M., Boer, H., & Houwen, R. J. (1989). Newspaper coverage of hazards and the reactions of readers. *Journalism Quarterly, 66*(4), 846–852.

Wiener, C. (1981). *The Politics of Alcoholism: Building an Arena around a Social Problem*. New Brunswick, NJ: Transaction.

Williams, A. (2015). Environmental news journalism, public relations and news sources. In A. Hansen & R. Cox (Eds.), *The Routledge Handbook of Environment and Communication* (pp. 197–205). London and New York: Routledge.

Williams, R. (1973). *The Country and the City*. London: Chatto & Windus.

Williams, R. (1983 [1976]). *Keywords: A Vocabulary of Culture and Society*. London: Flamingo/Fontana.

Williamson, J. (1978). *Decoding Advertisements: Ideology and Meaning in Advertising*. London: Marion Boyars.

Williamson, J. (2010). *Decoding Advertisements: Ideology and Meaning in Advertising* (Reissue ed.). London: Marion Boyars.

Wilson, A. (1992). *The Culture of Nature: North American Landscape from Disney to the Exxon Valdez*. Cambridge, MA: Blackwell.

Wozniak, A., Wessler, H., & Lück, J. (2016). Who prevails in the visual framing contest about the United Nations Climate Change Conferences? *Journalism Studies, Published online: 05 Feb 2016*, 1–20. doi: 10.1080/1461670X.2015.1131129

Xie, L. (2015). The story of two big chimneys: a frame analysis of climate change in US and Chinese newspapers. *Journal of Intercultural Communication Research, 44*(2), 151–177. doi: 10.1080/17475759.2015.1011593

Zhao, X., Leiserowitz, A. A., Maibach, E. W., & Roser-Renouf, C. (2011). Attention to science/environment news positively predicts and attention to political news negatively predicts global warming risk perceptions and policy support. *Journal of Communication, 61*(4), 713–731. doi: 10.1111/j.1460–2466.2011.01563.x

Zucker, H. G. (1978). The variable nature of news media influence. In B. D. Ruben (Ed.), *Communication Yearbook* (Vol. 2, pp. 225–240). New Brunswick, NJ: Transaction.

Index